Green Electricity
and Global Warming

Green Electricity
and Global Warming

RICHARD L. ITTEILAG

authorHOUSE®

AuthorHouse™
1663 Liberty Drive
Bloomington, IN 47403
www.authorhouse.com
Phone: 1-800-839-8640

© *2012 by Richard L. Itteilag. All rights reserved.*

No part of this book may be reproduced, stored in a retrieval system, or transmitted by any means without the written permission of the author.

Published by AuthorHouse   08/01/2012

ISBN: 978-1-4772-1741-2 (sc)
ISBN: 978-1-4772-1739-9 (hc)
ISBN: 978-1-4772-1740-5 (e)

Library of Congress Control Number: 2012910979

Any people depicted in stock imagery provided by Thinkstock are models, and such images are being used for illustrative purposes only.
Certain stock imagery © Thinkstock.

This book is printed on acid-free paper.

Because of the dynamic nature of the Internet, any web addresses or links contained in this book may have changed since publication and may no longer be valid. The views expressed in this work are solely those of the author and do not necessarily reflect the views of the publisher, and the publisher hereby disclaims any responsibility for them.

Contents

Preface vii
State Utility Demand Programs For Energy Efficiency 1
Utility Demand Programs 15
Estimated Costs Of Co2 Emissions Allowances 25
Co2 Emissions Trading Fundamentals 29
New Industrial Energy-Efficiency Technologies
    And The Effects On Industrial Energy Intensity
    (I.E., Industrial Conservation) *Part One* 103
New Industrial Energy-Efficiency Technologies
    And The Effects On Industrial Energy Intensity
    (I.E., Industrial Conservation) *Part Two* 109
Nuclear Power Generation:
    Capital, Operating And Enduser Costs 115
"Potential Conservation Created by National Demand
    Response Programs on Electricity Demand Capacity" 119
"Potential Conservation Created by Regional Demand
    Response Electricity Capacity" 123
EPA Co2 Emissions Regulations
    And Their Impact On Industrial Customers 143
The New Natural Gas: Drilling Technologies
    And U.S. Producing Zones 147
Utility Off-Peak Electricity Load Leveling Programs 149
Electricity Peak-Shaving Techniques 153

Preface

Electricity is the "power to succeed." However, the United States faces a hidden electricity crisis, i.e., the "power to fail." As the economy grows at 2-3 percent per year, the total demand for electricity has grown in tandem at a comparable 2-3 percent per year rate. Inversely, however, electricity capacity margins, the percentage of "spinning" supply above actual demand, have declined consistently over the last decade from 25-30 percent in 1992 to about 15 percent today. In fact, the Eastern independent power grid, with nearly 75 percent of total U.S. electricity demand, has only a 13.9 percent capacity margin.

To meet that level of demand growth and simultaneously offset the significant decline in capacity margins, the electric utility industry, for example, is planning 138 new coal-fired power plants nationally or the equivalent of $108 billion of new coal-generation capacity. In particular, the American Electric Power Company (AEP) has approved a major new electricity transmission line from West Virginia through Maryland and Pennsylvania to New Jersey in order to satisfy the largest electricity demand region in the country (PJM-Pennsylvania, New Jersey and Maryland).

Similarly, the Energy Policy Act of 2005 (EPact) grants up to $13 billion in subsidies to the nuclear industry to build new facilities, i.e., add new electricity supply. Entergy, one of the nation's largest utilities, has plans to add reactors to existing plants at Port Gibson and St. Francisville, near Baton Rouge, Louisiana. To date, there are plans for 16-19 new nuclear plants after plans for only three new reactors under consideration in 2005 and a national moratorium on the construction of new plants since 1978.

Richard L. Itteilag

This compound total demand growth coupled with declines in utility plant capacity margins only masks the serious underlying problem: peak electricity demand, typically for summertime air conditioning, is growing at 5 percent per year, considerably faster than total electricity demand. While the nation considers the need for energy independence as a critical energy policy initiative due to the fact that half the nation's oil consumption is imported and oil is the nation's primary transportation fuel, the resulting economic consequences of an electricity shortfall would be equally or more severe given the nation's near total reliance on electricity to cool, light and power motors and computers. That is the nexus of this book: electrical load-leveling or the techniques required to reduce electricity peaks in the U.S.

State Utility Demand Programs For Energy Efficiency

State legislative energy efficiency programs are widespread across the United States (U.S.). The programs range from financial incentives to program management. The most widely utilized are some form of program funding. The basis of this article is the available programs at utilities in five states to reduce electricity consumption or promote conservation in the U.S.

Alabama

State Loan Program: Local Government Energy Loan Program
Applicable Sectors: Schools, Local Government
Energy Efficiency Conservation Technologies: Heat Recovery Systems, Boiler efficiency and central plant improvements, upgrading water treatment plants for efficiency and/or recovery, HVAC equipment, etc . . .
Website: http://www.adeca.alabama.gov/C3/Local%20 Government%20Energy%20Loan%20P/default.aspx

California

Utility Rebate Program: City of Palo Alto Utilities-Commercial Advantage Energy Efficiency Program
Applicable Sectors: Commercial, Industrial
Energy Efficiency Conservation Technologies: City of Palo Alto is offering commercial / industrial customers incentives to replace old equipment with new and more efficient equipment such as steam traps for existing steam systems.

Website: http://www.cityofpaloalto.org/depts/utl/news/details.asp?NewsID=471&TargetID=223

Utility Rebate Program: PG&E-Nonresidential Energy Efficiency Rebates
Applicable Sectors: Commercial, Industrial, Nonprofit, Schools, Institutional, Agriculture
Energy Efficiency Conservation Technologies: Pacific Gas and Electric Company (PG&E) is offering rebates and incentives to businesses and nonresidential customers to increase energy efficiency. Specified improvements includes: Boilers and Water Heaters and HVAC improvements.
Website: http://www.pge.com/mybusiness/energysavingsrebates/rebatesincentives/ref/

Colorado

Utility Rebate Program: Longmont Power & Communications-Commercial Energy Efficiency Rebate Program
Applicable Sectors: Commercial, Industrial, Institutional
Energy Efficiency Conservation Technologies: Longmont Power & Communications, in collaboration w/ Platte River Power Authority, is offering an incentive for its business customer's incentive to install new or retrofitted energy efficient equipment in their facilities. There are no specified products and it is left customers to find the best way to save energy within their facilities.
Website: http://www.ci.longmont.co.us/lpc/bus/eep_homepage.htm

Utility Rebate Program: Loveland Water & Power-Commercial Energy Efficiency Rebate Program
Applicable Sectors: Commercial, Industrial
Energy Efficiency Conservation Technologies: Loveland Water & Power, in collaboration w/ Platte River Power Authority, is offering an incentive for its business customer's incentive to install new or retrofitted energy efficient equipment in their facilities. There are no specified products and it is left customers to find the best way to save energy within their facilities.

Website: http://www.ci.loveland.co.us/wp/power/conservation/c&i_ee.htm

Utility Rebate Program: Fort Collins Utilities-Commercial and Industrial Energy Efficiency Rebate Program
Applicable Sectors: Commercial, Industrial
Energy Efficiency Conservation Technologies: Fort Collins is providing commercial and industrial businesses incentives for new construction projects and existing building retrofits. Mechanical systems and industrial process improvements is encouraged by this program to assist facilities with new energy efficient equipment.
Website: http://www.fcgov.com/conservation/biz-eep.php

Connecticut

State Rebate Program: Connecticut Energy Efficiency Incentive Program
Applicable Sectors: Commercial, Industrial, Schools, Nonprofit, Local Government, Agricultural, Institutional
Energy Efficiency Conservation Technologies: Connecticut electricity customers that install energy efficiency equipment and reduce their energy use during peek hours may be eligible for a rebate.
Website: http://www.ameresco.com/ctdsmeep.asp

Utility Rebate Program: Connecticut Light & Power-Operation and Maintenance Program
Applicable Sectors: Commercial, Industrial
Energy Efficiency Conservation Technologies: Commercial and industrial customers are eligible for the Operations and Maintenance Program to help improve the maintenance and operation of electric equipment. Opportunities listed include improvements to compressed air systems, such as repairing leaks or installing solenoid valves that automatically shut-off air supply to machines when not running and repairs or replacement of defective steam traps.

Website: http://www.cl-p.com/business/saveenergy/default.aspx

Utility Rebate Program: The United Illuminating Company-Energy Opportunities Program
Applicable Sectors: Commercial, Industrial, Local, State and Federal Govt.
Energy Efficiency Conservation Technologies: Program is for retrofitting existing commercial, industrial or government buildings in an energy efficient manner. Includes Heat Recovery, Steam-system upgrades and process and manufacturing equipment. Energy / Engineering (Whole-System Concept) Studies are reimbursed after implementation up to $5K max to the customer.
Website: http://www.avistautilities.com/business/rebates/washington_idaho/Pages/incentive_10.aspx

Florida

Utility Rebate Program: Tampa Electric-Commercial Energy Efficiency Rebate Program
Applicable Sectors: Commercial, Industrial
Energy Efficiency Conservation Technologies: Tampa Electric is offering their customers a variety of incentives for the industrial and commercial sectors. Incentive is to increase the efficiency for their facilities, which includes heat recovery.
Website: http://www.tampaelectric.com/business/saveenergy/

Utility Rebate Program: Lakeland Electric-Commercial Conservation Rebate Program
Applicable Sectors: Commercial, Industrial
Energy Efficiency Conservation Technologies: Lakeland Electric will pay 50% of the cost, up to $2K, for their commercial customers to have an energy audit performed in their facilities. Please contact Lakeland Electric (863)834-9535 and ask about their PROFESSIONAL OUTSIDE ENERGY AUDIT PROGRAM.

Website: http://www.lakelandelectric.com/Business/
CommercialIndutrialInstitutional/ProductsServices/
EnergyEfficiencyPrograms/tabid/105/Default.aspx

Idaho

Utility Rebate Program: Idaho Power-Building Efficiency for Commercial Construction Rebate Program
Applicable Sectors: Commercial, Industrial
Energy Efficiency Conservation Technologies: Idaho Power is offering incentive for its industrial and commercial customers in Idaho and Oregon to upgrade to more efficient equipment in their facilities. Rebates include HVAC equipment and controls and other equipment. Customers interested must first submit a preliminary applications. Contact Customer Service (800)488-6151
Website: http://www.idahopower.com/EnergyEfficiency/Business/Programs/BuildingEfficiency/default.cfm

Illinois

State Rebate Program: DCEO-Large Customer Energy Analysis Program (LEAP)
Applicable Sectors: Commercial, Industrial, Schools, Local Govt and Hospitals
Energy Efficiency Conservation Technologies: Large energy users (w/ $500K+ energy cost) within the applicable sectors has the opportunity to receive an opportunity rebates of 50%, up to $10K towards the cost for developing an energy efficiency action plan and 50%, up to $10K, for the cost of a technical assessment.
Website: http://www.commerce.state.il.us/dceo/Bureaus/Energy_Recycling/Energy/

Utility Rebate Program: Nicor Gas Energy Efficiency Program for Business Customers
Applicable Sectors: Commercial, Industrial
Energy Efficiency Conservation Technologies: Nicor Gas Energy Efficiency Program is offering business customers on Rates 4 and 74, opportunities to save energy and money through efficiency rebates and energy-saving ideas. Steam Traps are one of the qualifying products that must be purchased and installed between May 1, 2010-May 31, 2011. Rebate for a Steam Trap Replacement is $50.00. Annual natural Gas Savings (Therms) $203 / Annual Cost Savings Estimate $132
Website: http://members.questline.com/Article.aspx?articleID=14480&accountID=1951&nl=7966&userID=973599

Indiana

Utility Rebate Program: Citizens Gas-Commercial Efficiency Rebates
Applicable Sectors: Commercial
Energy Efficiency Conservation Technologies: Citizens Gas of Indiana is offering rebates to its commercial customers for the installation of several different types of efficient natural gas equipment, which includes certain steam trap services
Website: http://www.citizensgas.com/forhomes/appliancerebates.html

Utility Rebate Program: Vectren Energy Delivery-Commercial Energy Efficiency Rebates
Applicable Sectors: Commercial, Industrial
Energy Efficiency Conservation Technologies: Vectren Energy Delivery is offering natural gas customers rebates for the installation of certain types of highly efficient natural gas equipment for technologies such as steam-system upgrades.
Website: https://www.vectrenenergy.com/web/eenablement/learn_about/conservation/rebates_i.jsp

Kentucky

Utility Rebate Program: Duke Energy-Commercial and Industrial Energy Efficiency Rebate Program
Applicable Sectors: Commercial, Industrial, Schools
Energy Efficiency Conservation Technologies: Duke Energy is offering incentives for manufacturing process equipment. The maximum incentive is up to $50K per fiscal year per facility. Applications for incentives must be completed and submitted within 90 days after equipment has been installed and is operational.
Website: http://www.duke-energy.com/

Maine

Utility Rebate Program: Northern Utilities-Commercial Energy Efficiency Programs
Applicable Sectors: Commercial, Industrial, Schools, Local Government, Institutional
Energy Efficiency Conservation Technologies: Program includes energy efficient upgrades with includes heat recovery equipment and systems.
Website: http://www.northernutilities.com/business/eneraudit.htm

Massachusetts

Utility Rebate Program: Bay State Gas-Commercial Energy Efficiency Program
Applicable Sectors: Commercial, Industrial
Energy Efficiency Conservation Technologies: Bay State Gas Co is offering its commercial and industrial customers rebates for energy efficient improvements to their facilities.
Website: http://www.baystategas.com/en/save-energy-money/Energy-Efficiency-Business.aspx

Utility Rebate Program: National Grid-Commercial (Electric) Energy Efficiency Incentive Program
Applicable Sectors: Commercial, Industrial, Schools, Local Government
Energy Efficiency Conservation Technologies: National Grid offers electric energy efficiency programs for its larger commercial and industrial customers. An Energy Efficiency Engineering Study must be completed by a certified energy manager (CEM) or a professional engineer (PE) with a completed application submitted for approval by Dec 31 2009. Acceptable study such as heat recovery of process loads.
Website: http://www.thinksmartthinkgreen.com/

Utility Rebate Program: National Grid-Commercial (Gas) Energy Efficiency Incentive Program
Applicable Sectors: Commercial
Energy Efficiency Conservation Technologies: National Grid's Commercial Energy Efficiency Incentive Programs provides support services to commercial customers who install energy efficient natural gas related measures. Prescriptive rebates are available for common energy efficiency measures installed after the completion of an energy audit, including steam trap replacements.
Website: http://www.thinksmartthinkgreen.com/

Michigan

Utility Rebate Program: Consumers Energy-Commercial Energy Efficiency Program
Applicable Sectors: Commercial, Industrial
Energy Efficiency Conservation Technologies: Incentives available from Consumers Energy for energy efficiency equipment upgrades and improvements, which includes steam-system upgrades. Program expires when approved funds are exhausted or until December 31st of each program year reaching year 2014.
Website: http://www.consumersenergy.com/eeprograms/Landing.aspx?ID=750

Utility Rebate Program: DTE Energy-Commercial Energy Efficiency Program
Applicable Sectors: Commercial, Industrial, Local Government
Energy Efficiency Conservation Technologies: Incentives available from DTE Energy for qualified energy measures. Steam-system upgrades is one of the programs eligible efficiency technology. For more details visit the below website or contact DTE Energy Efficiency (866) 796-0512
Website: http://www.YourEnergySavings.com

Utility Rebate Program: Michigan Gas Utilities (Efficiency United)-Commercial and Industrial Rebate Program
Applicable Sectors: Commercial, Industrial
Energy Efficiency Conservation Technologies: Various rebates and incentives to commercial and industrial gas customers of Michigan Gas Utilities (Efficiency United) for qualified energy equipment and measures. Steam-system upgrades is one of the programs eligible efficiency technology. Projects must be implemented by December 31, 2010. Visit the program web site for guidelines and application.
Website: http://www.efficiencyunited.com/util_michgasutils.asp

Utility Rebate Program: SEMCO Energy (Efficiency United)-Commercial and Industrial Rebate Program
Applicable Sectors: Commercial, Industrial
Energy Efficiency Conservation Technologies: Program offers rebates and incentives to commercial and industrial gas customers of SEMCO Energy service area. Eligible equipment and measures includes steam trap tests and replacements, boiler tune-ups and controls, etc . . . Contact SEMCO Energy for approval before work / purchases begin. SEMCO Energy Gas Company PH: (800) 624-2019 / Efficiency United PH: (877) 367-3191. Applicants have 120 days to implement project after approval is granted. Applications can be found on the website. Projects must be implemented by December 31, 2010.
Website: http://www.efficiencyunited.com/util_semco.asp

Utility Rebate Program: DTE Energy-Commercial Energy Efficiency Program
Applicable Sectors: Commercial, Industrial, Local Government
Energy Efficiency Conservation Technologies: Incentives available from DTE Energy for qualified energy measures. Steam-system upgrades is one of the programs eligible efficiency technology. For more details visit the below website or contact DTE Energy Efficiency (866) 796-0512
Website: http://www.YourEnergySavings.com

Minnesota

Utility Rebate Program: CenterPoint Energy in Minnesota-Boiler System Rebate Program
Applicable Sectors: Commercial
Energy Efficiency Conservation Technologies: Center Point Energy is helping their customers facilities operate their systems more efficiently. Included in their rebate program for Boilers are steam trap replacements. Rebate amount is 35% of equipment or up to $10K per building capital. Approval and installation must be complete before December 15th, 2009. Contact CenterPoint Energy for additional details.
Website: http://www.centerpointenergy.com/services/naturalgas/business/rebatesforbusiness/boilersystems/MN/

Missouri

Utility Rebate Program: AmerenUE-Commercial Natural Gas Equipment Rebates
Applicable Sectors: Commercial
Energy Efficiency Conservation Technologies: AmerenUE offers its commercial natural gas customer multiple rebates for the installation of certain energy efficient natural gas equipment. Steam Trap Replacements: $50/unit. Maximum incentive for steam trap replacements: $2500 or 50% of cost for up to 50 units. Equipment must be purchased by Dec 31, 2009 and

installed by Jan 31, 2010. Contact Ameren @ 800-210-8131 prior to making equipment purchases.
Website: http://www.ameren.com/NaturalGasCenter/ADC_2009RebateHome.asp

Utility Rebate Program: Empire District Electric-Commercial and Industrial Efficiency Rebates
Applicable Sectors: Commercial, Industrial, Nonprofit, Schools, Institutional
Energy Efficiency Conservation Technologies: Empire District Electric Co. is offering custom rebates for retrofits to certain commercial and industrial customers. Customers who plan to replace equipment in an existing facility with high-efficiency equipment can obtain a custom rebate for a retrofit project. Individual efficiency measures can also be approved by Empire District Electric for measures and equipment needs for your company. Contact Empire District Electric for pre-approval and additional details.
Website: http://empire.programprocessing.com/content/Home

New Hampshire

Utility Rebate Program: National Grid-Commercial Energy Efficiency Incentive Program
Applicable Sectors: Commercial
Energy Efficiency Conservation Technologies: National Grid's Commercial Energy Efficiency Incentive Programs provides support services and incentives to commercial customers who install energy efficient natural gas related measures. Prescriptive rebates are available for common energy efficiency measures installed after the completion of an energy audit, including steam trap replacements. The Commercial High Efficiency Heating Program offers rebates ranging from 100 to $6K for various types of energy efficient space and water heating equipment.
Website: http://www.thinksmartthinkgreen.com/
Utility Rebate Program: PSNH-Large Commercial and Industrial Energy Efficiency Rebate Program

Applicable Sectors: Commercial, Industrial

Energy Efficiency Conservation Technologies: Public Service of New Hampshire (PSNH) encourages its large commercial and industrial customers to conserve energy through their New Equipment and Construction Program and their Large Business Retrofit Program. Large Business Retrofit Program Custom Program to show Annual KWH Savings, Demand Savings (KW) and Dollar Savings. Program includes equipment, material and labor costs.

Website: http://www.psnh.com/Business/Efficiency/default.asp

Utility Rebate Program: PSNH-Unitil-Commercial and Industrial Energy Efficiency Programs

Applicable Sectors: Commercial, Industrial, Institutional

Energy Efficiency Conservation Technologies: Unitil offers three different programs: Small Business Energy Efficiency Program, Large C&I Retrofit Program & Large C&I New Construction Program. Custom Process Projects fall within Large C&I Retrofit Program "Custom Project" & Large C&I New Construction Program. Application must be submitted and approved with a savings report.

Website: http://services.unitil.com/nh/bus_energy_efficiency_programs.asp?t=6

New York

Utility Rebate Program: National Grid-Commercial (Gas) Energy Efficiency Incentive Program

Applicable Sectors: Commercial

Energy Efficiency Conservation Technologies: National Grid's Commercial (Gas) Energy Efficiency Incentive Programs provides eligible efficiency technologies including steam trap replacements and heat recovery.

Website: http://www.thinksmartthinkgreen.com/

North Carolina

State Rebate Program: Steam Trap Rebate Program
Applicable Sectors: Commercial, Industrial, Local Government, Institutional
Energy Efficiency Conservation Technologies: Steam Trap Survey Rebate Program provides steam trap survey services to facilities that use steam for heating or processing. A participating survey firms identifies steam traps and tests to see that they are working properly. Program participants will receive a set amount of funding for each steam trap surveyed.
Website: http://www.energync.net/programs/industry.html

Oregon

Utility Rebate Program: Springfield Utility Board-Energy Savings Plan Program
Applicable Sectors: Industrial
Energy Efficiency Conservation Technologies: The Springfield Utility Board provides industrial customers with a comprehensive report to identify cost effective efficiency improvements. Eligible measures includes process modification, energy mgmt control systems, cooling tower conversions, heat recovery equipment, etc.
Website: http://www.subutil.com/conservation_services/for_your_business

Utility Rebate Program: Idaho Power-Building Efficiency for Commercial Construction Rebate Program
Applicable Sectors: Commercial, Industrial
Energy Efficiency Conservation Technologies: Idaho Power is offering incentive for its industrial and commercial customers in Idaho and Oregon to upgrade to more efficient equipment in their facilities. Rebates include HVAC equipment and controls and other equipment. Customers interested must first submit a preliminary applications. Contact Customer Service (800)488-6151
Website: http://www.idahopower.com/EnergyEfficiency/Business/Programs/BuildingEfficiency/default.cfm

Rhode Island

Utility Rebate Program: National Grid-Commercial (Gas) Energy Efficiency Programs
Applicable Sectors: Commercial
Energy Efficiency Conservation Technologies: Program provides services and incentives to commercial customers who install energy efficient natural gas related measures. Prescriptive rebates are available for common energy measures installed after the completion of an energy audit, such as: steam trap replacements.
Website: http://www.thinksmartthinkgreen.com/

Washington

Utility Rebate Program: Clark Public Utilities-Commercial Energy Efficiency Rebate Program
Applicable Sectors: Commercial, Industrial
Energy Efficiency Conservation Technologies: Clark Public Utilities offers a variety of energy efficiency rebates and services to help their commercial and industrial customers. For commercial and industrial customers that are replacing existing equipment or installing new equipment that is energy efficient, the utility offers the custom project program. Examples of eligible project include: process energy system improvements. All proposals must be measurable and verifiable.
Website: http://www.clarkpublicutilities.com/yourbusiness/businessConservationPrograms/

Today, utility demand programs are widespread across the U.S. The programs range from financial incentives to program management. The most widely utilized are some form of program funding. Generally, most states require these types of demand programs. New York and California are the two states with the requirements for demand programs across all the utilities in those states.

9/30/10

UTILITY DEMAND PROGRAMS

BY

RICHARD L. ITTEILAG
PRESIDENT
ENERGISTICS, INC.
A DIVISION OF WASHINGTON
PROPOSALASSOCIATES, INC.

Abstract

Utility demand programs are widespread across the United States (U.S.). The programs range from financial incentives to program management. The most widely utilized are some form of program funding. That is the basis of this article: the available demand programs at utilities to reduce electricity consumption or promote conservation in the U.S.

Discussion

A recent report by the **U. S. Environmental Protection Agency, October 2008 Combined Heat and Power Partnership** entitled: **UTILITY INCENTIVES FOR COMBINED HEAT AND POWER** studied the numerous demand programs offered by utilities across the U.S. The programs ranged from financial incentives to program management.

Far more common are types of activities that promote combined heat and power (CHP) development. This report identified 16 types of utility actions/programs that support (CHP) development in ways other than offering direct financial incentives for (CHP) system installation:
• Program Funding (e.g., system benefits charge)
  Financial Incentives
  State Program Funding
   • Request for Proposals (RFP) for Supply
   • CHP Research and Development (R&D)/Demonstration Projects
   • Outreach (e.g., CHP-specific Web page)
   • Site/Feasibility Analyses
   • Design and Engineering
   • Construction and Installation
	• Maintenance and Operation
	• Project Management
	• Ownership/Joint Ownership
	• Performance Contracting
	• Favorable Gas Rates
	• Load Curtailment Payments
	• Regulatory Process Advice
	• Shared Savings Loans (waste heat recovery)
	• Custom Rebates (waste heat recovery)

Due to increasing energy costs, expanding load growth and state/local initiatives to decrease energy consumption and lower carbon emissions, there is increased utility interest in (CHP) across the United States. These (CHP) projects can yield numerous benefits to electric and gas utilities and to the public, including:

- Bringing economic development to a state.
- Reducing peak electrical demand on the grid.
- Yielding improvements to electric grid system efficiency by reducing grid congestion.
- Deferring or displacing more expensive transmission and distribution infrastructure investments.
- Reducing the environmental impact of power generation.
- Helping to meet state mandated renewable portfolio standards in states where (CHP) constitutes an eligible resource.
- Reducing fuel price volatility.

Utilities have also been involved to varying degrees in state-initiated programs to promote (CHP). These state programs typically funded by system benefits charges (SBCs) levied on customer bills—have resulted in hundreds of operational (CHP) projects and nearly 200 megawatts (MW) of supply. Utility actions in response to state policies and initiatives to promote (CHP) range from collecting mandated system benefits charges via customers' gas and electric bills (e.g., in New York, Vermont), to assisting customers with accessing funding available from the state (e.g., in Connecticut, New Jersey), to administering state-initiated programs (e.g., in California, Minnesota).

The most widely utilized are some form of program funding. Generally, most states require these types of demand programs. New York and California are the two states with the requirements for demand programs across all the utilities in those states. In New York, Consolidated Edison Company has the most effective demand program: load curtailment payments to promote conservation. The California utilities all offer a web page to promote (CHP). The table below lists all the demand programs offered in the U.S.:

Financial Incentives
State Program Funding
RFP for Supply
CHP R&D/ Demonstration
Outreach
CHP Web page
Site /Feasibility
Design & Engineering
Construction & Installation
Maintenance & Operation
Project Management
Ownership/Joint Ownership
Performance Contracting
Favorable Gas Rates
Load Curtailment Payment
Regulatory Process Advice
Shared Savings Loans (WHR)
Custom Rebates (WHR)

Conclusion

Today, utility demand programs are widespread across the U.S. The programs range from financial incentives to program management. The most widely utilized are some form of program funding. Generally, most states require these types of demand programs. New York and California are the two states with the requirements for demand programs across all the utilities in those states.

Agriculture's Impact on Global Warming (Nov. 2010)

As we struggle to grow enough food to feed the planet, we are clearing forests that are a crucial protection against the warming climate, UW-Madison researchers have found.

Scientists, for the first time, analyzed the tradeoff between agricultural production and the capacity of forests and other natural ecosystems to store carbon. Without the storage capacity of forests, more carbon dioxide—the gas that is causing the climate to warm—is released into the atmosphere.

The research is important, according to UW-Madison scientist Paul West, because it could lead to practical methods of balancing our need to grow food and efforts to slow or counter climate change.

The study was published online Monday in the Proceedings of the National Academy of Sciences. It's part of a special look by the journal at climate and agricultural productivity in the tropics. The research team also included scientists the University of Minnesota, Stanford University, Arizona State University and The Nature Conservancy.

West, a UW-Madison graduate student and the lead author of the study, said much of the work was focused on the tropics because of the tremendous amount of carbon stored in the trees and soil there—more than twice that of temperate forests. At the same time, increasing populations and concerns about food supply are driving an increase in the tropical acreage devoted to agriculture and a decrease in forested lands.

Using satellite data and government reports, the researchers studied the cultivation of 175 different crop plants around the world. They compared that information to estimates of the amount of carbon stored in natural vegetation, such as prairie grass or tropical trees. They then created a map that compares crop yield with carbon loss.

For example, calculations showed that, for every ton of crop yield in the tropics, carbon stocks are decreased by as much as 75 tons.

That tradeoff is especially suspect in the tropics, West said, because crop yields are not as high as in temperate regions.

Jonathon Foley, formerly a researcher at UW-Madison and now head of the Institute on the Environment at the University of Minnesota, said the study offers data that has practical implications. It may be motivation, for example, for finding ways to make food production around the world more efficient and reducing the acres cleared for farming.

"We grow a lot of food that is wasted," Foley said.

The difficulty, West said, is balancing the great need for food with the necessity of forestalling the potentially devastating impacts of climate change. "It's one of our grand challenges," West said.

Climate Change (Oct. 2010)

Candidates in 2010 who assume that Tea Party supporters and independents will respond to the same messages on climate and clean energy issues appear to be mistaken, according to a major new survey of more than 1,000 Americans conducted by Opinion Research Corporation for the nonprofit and nonpartisan Civil Society Institute (CSI). Further, while the views of Americans on climate science issues are now divided sharply along partisan lines, there remains strong support for "concrete" action focused on protecting clean air and clean water.

This poll provides one more data point that while independents have swung largely to supporting Republicans in the November 2 elections, they are not following tea party activists in lockstep.

One of the key Republican messages of the current campaign has been a theme opposing any new federal regulations, such as those which could reduce greenhouse gas emissions. One of the largest funders of attack ads against Democrats this year has been the U.S. Chamber of Commerce, a highly pro-business group which opposes climate legislation.

Although enactment of legislation to curb the emissions that cause climate change has been a top priority for President Obama, Congress has failed to send him a bill to sign. The House approved cap-and-trade legislation in the summer of 2009, but a similar measure stalled in the Senate, failing to capture any GOP support.

Key findings of the new CSI poll include the following:

- 2 percent of Americans say they are "an active member of the Tea Party movement," 23 percent support the Tea Party, 36 percent have no view about the Tea Party, and 28 percent oppose the Tea Party.
- Independents are more than twice as likely as Tea Party supporters (62 percent versus 27 percent) to see global

warming as a problem in need of a solution, compared to 39 percent of Republicans and 82 percent of Democrats. Overall, more than three out of five Americans agree that "(g)lobal warming and climate change are already a big problem and we should be leading the world in solutions," compared to about a quarter (27 percent) who think "(g)lobal warming may or may not be happening. We should let other countries act first while the science sorts itself out."

- Tea Party supporters are more than twice as likely as Independents (34 percent versus 15 percent) to see no need for leadership on global warming, compared to 29 percent of Republicans and 8 percent of Democrats. Overall, only 17 percent of Americans see no need for "national OR grassroots leadership on global warming." Another 12 percent think no federal leadership on energy policy is needed "since some grassroots officials are taking actions," compared to 61 percent who think "(w)e need leadership on energy policy from Washington, D.C., because it is a national problem that will require national solutions."
- However, the partisan divide is far less sharp when the discussion turns to specifics. Just over three out of four (76 percent) Americans think that—when it comes to energy sources, such as natural gas, coal, tar sands, nuclear and biofuels, requiring a high amount of water for production purposes—"(w)ater shortages and clean drinking water are real concerns. America should put the emphasis on first developing new energy sources that require the least water and have minimal water pollution." Only 13 percent agreed with this statement: "Energy supply needs should override concerns about water shortages and water pollution. America should proceed first with developing energy sources even if they may have significant water pollution and water shortage downsides." Supporters of putting the primary emphasis on clean water include 68 percent of Republicans, 80 percent of independents, 81 percent of Democrats and 60 percent of Tea Party supporters.

"These findings point to a greater diversity of views among Tea Party supporters and independents than is widely assumed to be the case, and this has major implications for the 2010 elections and future elections," says Opinion Research Corporation senior researcher Graham Hueber. "What we are seeing here is a common mistake with which pollsters are all too familiar: the tendency on the part of the media and others to simplify the story by lumping together groups rather than being careful to parse out the specific points on which they actually differ and sometimes quite dramatically so."

Melody Barnes, President Obama's top domestic policy adviser, said today that the Obama administration does not have second thoughts about pushing health care legislation before a cap-and-trade bill. Some have argues that the administration's aggressive push to pass a health care bill took the wind out of the sails of efforts to pass a climate bill in the Senate.

Barnes, speaking at The Atlantic's Green Intelligence Forum today, said that health care reform was one of Obama's top priorities during the campaign. "One of the things that we heard is that we had to deal with the issue of health care," she said.

Barnes also demurred on the question of a climate and energy bill's legislative prospects. "One of the things that I've learned is that when you start to put your money down on when Congress will act, you're going to lose your money," she said. But she stressed that while Congress was far from "crossing the finish line" on a climate bill, the Obama administration is committed to using its regulatory authority, at the Environmental Protection Agency and elsewhere, to address climate change.

Barnes expressed frustration with those who oppose measures to address climate change. "There is no debate globally about the importance of this issue, it's when we return home that we are pushing and shoving and trying to convince people of the importance of this issue," she said. But she added that she is seeing a shift in the viewpoint of the American people on climate change, though she acknowledged that many Americans still question the science of climate change.

"Even if you, for whatever reason, don't believe the science, you've got to believe the economic imperative of a clean energy economy," she said.

7/28/10

ESTIMATED COSTS OF CO2 EMISSIONS ALLOWANCES

BY

RICHARD L. ITTEILAG
PRESIDENT
ENERGISTICS, INC.
A DIVISION OF WASHINGTON PROPOSAL
ASSOCIATES, INC.

Abstract

While the United States (U.S.) is presently debating **Emissions trading,** also known as '**cap and trade'**, Europe has a plan in effect at present. The European program uses carbon dioxide (CO_2) limits and economic incentives to administer its plan. The U.S. is expected to use this plan as a guide to future legislation. This paper discusses the cost estimates for CO_2 allowances of a 'cap and trade' policy.

Discussion

As a candidate, Barack Obama said he'd tackle climate change by imposing caps on emissions of greenhouse gases. Now, as President, he's doing exactly that. He proposes reducing U.S. emissions 14% below 2005 levels by 2020 and 83% below by 2050. And he'd raise $646 billion from 2012 to 2019 by auctioning the rights to emit such gases—in effect putting a price on carbon emissions. With Congress also serious about the climate, business knows the battle has been joined for real and is trying to shape a compromise bill likely to emerge this year. "We are now playing with live bullets," says the Environmental Defense Fund's Mark Brownstein, who works with a group of companies that supports the plan.

The bullets are already flying—but mainly over details of the plan, not the general idea. While there are still fierce opponents of emissions limits, such as the U.S. Chamber of Commerce, much of business is supportive. The Obama Administration "is very close to right on the climate plan," says John W. Rowe, chief executive of Exelon, a Chicago-based utility.

In theory, a workable cap-and-trade market for carbon emissions would give business executives more certainty about future energy costs, helping them make better investment decisions. A market price on carbon would boost energy efficiency and renewable energy efforts, already beneficiaries in Obama's stimulus package. Nuclear power plants, such as Exelon's, would become more valuable. "I have great hope for the 'green' stimulus, but it won't fulfill its potential unless there is a price on carbon," says James E. Rogers, chief executive of Duke Energy. Also, there's little chance of getting China and India to agree to binding limits, which American companies insist is needed to keep the international playing field level, unless the U.S. takes action at home.

The real fight, therefore, is not whether to impose carbon limits but how to do so and at what cost to business. Obama proposes that companies buy an allowance, or permit, for each ton of carbon emitted, at an estimated cost, to start, of $13 to $20 per ton. (Those permits could also be bought and sold.) Even at the lower range of $13 per ton, energy companies and utilities would likely pass along the added cost to consumers. It's estimated the price of gasoline would go up by 12 cents a gallon and the average electricity bill by about 7% nationally—and far higher in states more dependent on coal. Unfair, say many executives. "It is a clear transfer of the middle part of the country's wealth to the two coasts," says Michael G. Morris, CEO of American Electric Power, a coal-heavy power generator based in Columbus, Ohio, that supplies electricity in 11 states.

Morris intends to target the 50 U.S. senators in the 25 coal-centric states "to see if we can bring some rationality to the program," he says. The U.S. Chamber of Commerce, meanwhile, plans to hold "climate dialogues" in as many as 16 cities, hammering home a similar message in coal-rich states with Democratic senators. The Obama plan "is now a very expensive tax used to transfer wealth. It has nothing to do with climate change," charges William L. Kovacs, a Chamber vice-president.

The Obama team points out that its cap-and-trade plan returns much of the money raised by permit sales to consumers nationwide in the form of lower taxes, so many people come out ahead. And the Environmental Defense Fund has created a map of 1,200 alternative energy or energy-efficiency companies in key manufacturing states that stand to benefit from the climate plan. While the Midwest will bear a higher cost from reducing carbon emissions, the region will also benefit from the most new jobs, the EDF argues.

Lots of other details remain to fight over. Dow Chemical and others want credit for emission cuts they have already made, for example. So prepare for months of negotiations. But a deal is likely. Says Dow lobbyist Peter A. Molinaro: "Somewhere out there is a rational policy that could actually get the votes."

Conclusion

The U.S. industrial sector will likely see climate change regulations, i.e., CO_2 emissions caps, within the next three years. These regulations will either be legislated by the Congress, or at a minimum, be mandated by the EPA. Obama proposes that companies buy an allowance, or permit, for each ton of carbon emitted, at an estimated cost, to start, of $13 to $20 per ton. The fundamentals of the plan will most likely mirror a plan in place in Europe presently and be a 'cap and trade' policy administered by the U.S. government with CO_2 limits and economic incentives.

5/31/10

CO2 EMISSIONS TRADING FUNDAMENTALS

BY

RICHARD L. ITTEILAG
PRESIDENT
ENERGISTICS, INC.
A DIVISION OF WASHINGTON PROPOSAL
ASSOCIATES, INC.

Abstract

Emissions trading, also known as **'cap and trade'**, is an administrative approach used to control pollution. While the United States (U.S.) is presently debating this issue, Europe has a plan in effect at present. The European program uses carbon dioxide (CO_2) limits and economic incentives to administer its plan. The U.S. is expected to use this plan as a guide to future legislation. This paper discusses the cost estimates for CO_2. allowances of a 'cap and trade' policy.

Discussion

As a candidate, Barack Obama said he'd tackle climate change by imposing caps on emissions of greenhouse gases. Now, as President, he's doing exactly that. He proposes reducing U.S. emissions 14% below 2005 levels by 2020 and 83% below by 2050. And he'd raise $646 billion from 2012 to 2019 by auctioning the rights to emit such gases—in effect putting a price on carbon emissions. With Congress also serious about the climate, business knows the battle has been joined for real and is trying to shape a compromise bill likely to emerge this year. "We are now playing with live bullets," says the Environmental Defense Fund's Mark Brownstein, who works with a group of companies that supports the plan.

The bullets are already flying—but mainly over details of the plan, not the general idea. While there are still fierce opponents of emissions limits, such as the U.S. Chamber of Commerce, much of business is supportive. The Obama Administration "is very close to right on the climate plan," says John W. Rowe, chief executive of Exelon, a Chicago-based utility.

In theory, a workable cap-and-trade market for carbon emissions would give business executives more certainty about future energy costs, helping them make better investment decisions. A market price on carbon would boost energy efficiency and renewable energy efforts, already beneficiaries in Obama's stimulus package. Nuclear power plants, such as Exelon's, would become more valuable. "I have great hope for the 'green' stimulus, but it won't fulfill its potential unless there is a price on carbon," says James E. Rogers, chief executive of Duke Energy. Also, there's little chance of getting China and India to agree to binding limits, which American companies insist is needed to keep the international playing field level, unless the U.S. takes action at home.

The real fight, therefore, is not whether to impose carbon limits but how to do so and at what cost to business. Obama proposes that companies buy an allowance, or permit, for each ton of carbon emitted, at an estimated cost, to start, of $13 to $20 per ton. (Those permits could also be bought and sold.) Even at the lower range of $13 per ton, energy companies and utilities would likely pass along the added cost to consumers. It's estimated the price of gasoline would go up by 12 cents a gallon and the average electricity bill by about 7% nationally—and far higher in states more dependent on coal. Unfair, say many executives. "It is a clear transfer of the middle part of the country's wealth to the two coasts," says Michael

G. Morris, CEO of American Electric Power, a coal-heavy power generator based in Columbus, Ohio, that supplies electricity in 11 states.

Morris intends to target the 50 U.S. senators in the 25 coal-centric states "to see if we can bring some rationality to the program," he says. The U.S. Chamber of Commerce, meanwhile, plans to hold "climate dialogues" in as many as 16 cities, hammering home a similar message in coal-rich states with Democratic senators. The Obama plan "is now a very expensive tax used to transfer wealth. It has nothing to do with climate change," charges William L. Kovacs, a Chamber vice-president.

The Obama team points out that its cap-and-trade plan returns much of the money raised by permit sales to consumers nationwide in the form of lower taxes, so many people come out ahead. And the Environmental Defense Fund has created a map of 1,200 alternative energy or energy-efficiency companies in key manufacturing states that stand to benefit from the climate plan. While the Midwest will bear a higher cost from reducing carbon emissions, the region will also benefit from the most new jobs, the EDF argues.

Lots of other details remain to fight over. Dow Chemical and others want credit for emission cuts they have already made, for example. So prepare for months of negotiations. But a deal is likely. Says Dow lobbyist Peter A. Molinaro: "Somewhere out there is a rational policy that could actually get the votes."

Conclusion

The U.S. industrial sector will likely see climate change regulations, i.e., CO_2 emissions caps, within the next three years. These regulations will either be legislated by the Congress, or at a minimum, be mandated by the EPA. The fundamentals of the plan will most likely mirror a plan in place in Europe presently and be a 'cap and trade' policy administered by the U.S. government with CO_2 limits and economic incentives.

Richard L. Itteilag

Consumers Energy Efficiency Successes (Nov. 2010)

Motivating customers and organizations to change their behavior can lead to significant energy savings, according to case studies published by the American Council for an Energy-Efficient Economy (ACEEE).

The report profiles a variety of programs that spur individuals and organizations to save energy by changing behavior in their homes, businesses, and plants. The selected case studies illustrate the results possible when applying social science to energy efficiency and conservation programs across the spectrum of customer types and the different ways they use energy.

"Helping consumers better understand their energy use, and providing them the knowledge and tools necessary to change the way they use energy are essential to achieving the full economic and environmental benefits from energy efficiency," said Dan York, Utilities Program Deputy Director.

Social science-based programs that seek to reduce customer energy use are attracting increased interest as governments, industries, and the public expand their energy efficiency efforts to accomplish environmental, economic, organizational, and personal goals. Key factors for success from these leading case studies include making energy use "visible" to customers, setting measurable goals, providing incentives and instructions for action, and providing feedback on progress towards customer goals.

This report features profiles of 10 large, recent programs that have met a broad range of efficiency targets—from 2% to 20% of participants' energy use—using a variety of approaches. "The case studies profiled in the report really show that behavioral programs can be incredibly effective, even outside of the buildings sector, where they are most commonly applied through utility programs," noted Jennifer Amann, Buildings Program Director and co-author of the report. "Our report documents the impact of behavior change on energy use in the transportation and industrial sectors for the first time."

"Our report discusses how employee and management programs create a corporate culture that achieves large energy savings while improving the profitability and competitiveness of manufacturing," said Neal Elliott, Associate Director for Research. "You can't manage what you don't measure."

Two of the case studies highlight the ambitious efforts by Alcoa (NYSE: AA) and Dow (NYSE: DOW) to reduce their total energy footprint in their companies through management and employee-led initiatives.

"The transportation sector presents a diverse set of opportunities for testing behavioral approaches. Smarter driving habits, reduced vehicle miles traveled, and efficient vehicle purchases can have a tremendous impact on fuel consumption and greenhouse gas emissions in personal vehicles and commercial fleets," said Shruti Vaidyanathan, Transportation Research Associate and co-author of the report.

Case studies from the transportation sector include EPA's SmartWay Transport Partnership as well as France's feebate program.

According to ACEEE, the report's case studies demonstrate that behavioral change by individual consumers can lead to significant energy and cost savings. Applying social science to energy efficiency and conservation programs can effectively guide individuals and organizations towards increased energy efficiency and reduced energy use.

Current GOP Legislative Agenda

With midterm elections complete and a GOP House in place, the newcomers have the opportunity to either push forward or detain renewable energy policy.

The new Congress has several ideas to work with that were held over from Obama's first two years in office, including a renewable electricity standard that expands the definition of what energy types count toward the thresholds, particularly adding emission-free nuclear power and clean coal to the definition of renewable energy. Some wonder if the new Republican Congress will be staunch policy conservatives like many of their predecessors, or flexible enough to implement changes that will positively affect renewable energy.

Tax incentives are sure to be a major driver to renewable energy development during this half-term. Rep. Dave Camp (R-MI) is in line to become the new chairman of the House Ways and Means Committee, a role that has significant influence over tax measures for renewables.

Camp is the current top Republican on the Ways and Means Committee. His Michigan District 4 is home to a number of wind and solar manufacturers, such as Dow Chemical, Dow Corning and United Solar Ovonic. Though a self-proclaimed supporter of alternative energy, Camp believes "it takes today's energy to power tomorrow's technology," as he said in the April 14, 2010 Hearing on Energy Tax Incentives Driving the Green Job Economy.

"You cannot increase the cost of producing 85 percent of the energy being used today and expect consumers or employers to benefit from tax incentives that are going to less than 10 percent of the energy being used today," Camp said.

Camp was referencing the fact that there was little change in America's reliance on fossil fuels from 2007 to 2009 despite the investment of nearly $40 billion in tax subsidies for renewables enacted in October of 2008. In the remarks Camp made during the April 14 hearing, he said that in 2007, petroleum, coal, nuclear and natural gas supplied 93 percent of America's energy, while renewable energy supplied only 7 percent. In 2009, 92 percent of the nation's energy came from petroleum, coal, nuclear and natural gas and 8 percent from renewables.

While Camp's words may seem discouraging to some renewable developers, his track record displays more openness toward renewables than some of his Republican predecessors. In October 2009, he cosponsored legislation to invest $2.25 billion for a solar technology research and development program and to create a committee to study the near and long-term research and development needs in solar technology. In November of 2009, he cosponsored legislation to amend the Internal Revenue Code of 1986 to allow an investment credit for property used to fabricate solar energy property. And as a push for the bulk of the power industry, Camp cosponsored legislation in January 2009 to amend the Clean Air Act to provide that greenhouse gases are not subject to the Act.

Elias Hinckley, a partner at the law firm Venable and professor of international energy policy at the Edmund A. Walsh School of Foreign Service at Georgetown University, said Camp's reputation as a proponent of clean energy could be beneficial to renewables. However, if Tea Party candidates instigate an energy tax, "that may guide some of his policy driving."

In regards to policy under the new House, Hinckley said the extension of existing tax subsidies for renewables is "relatively safe." However, the Treasury Grant Program may not be met with an extension. "I see some real difficulty considering its eroding support," Hinckley said.

Bob Cleaves, president and CEO of the Biomass Power Association, said Camp has been a supporter of biomass in the past and comes from a state that ranks in the top five states in the country in terms of biomass plants. "If the past is any judge of it, we anticipate continuing to getting a lot of support from within the Ways and Means Committee."

Cleaves said extension of the Treasury Grant Program and other subsidies will depend largely on what happens during the lame-duck session. "The question is: is there going to be enough time in a lame duck for Congress to address the extensions? We're very hopeful that it gets addressed before year-end."

Instead of the new House focusing on climate issues, Cleaves said he expects a greater emphasis on tax policy. "Whether a newly reconstituted Congress can get their arms around the idea of a federal Renewable Portfolio Standard, I think that remains to be seen."

Many political analysts are projecting the new House to offer suggestions for compromise, including tax breaks and incentives for investment in nuclear power, clean coal and renewable energy. Scott Segal, an industry lobbyist, told Politico that he expects Republicans to accept incentives for energy efficiency, nuclear power and hydroelectric power, coupled with credits for geothermal heat pumps and next-generation heating, ventilating and air conditioning.

"An approach like this would be very consistent with the expressed desire of the President to continue to focus on energy but to do so in 'chunks' as opposed to a comprehensive bill," Segal said.

Camp's philosophy seems to mesh with this expectation for new technology that will lessen reliance on foreign oil and encourage innovation through alternative and renewable fuels. In an issue statement, Camp said, "It is imperative that policies are in place to encourage the research and development of new, cellulosic fuels that use crop and animal waste and greater use of solar, wind, clean coal and other new energy technologies."

Customer Generation

As an SCE customer, you or your business may want to generate your own power to supplement the electricity you purchase from SCE. "Self-generation," also called "distributed generation," can serve various purposes that include:

- "Back-up" or emergency generation designed to be used during utility power outages
- "Cogeneration," or combined heat and power applications, used by customers that have consistently high need for steam or another form of thermal energy
- Generation to be used during "peak demand," when it may be less costly to operate a generator than to buy power from SCE
- "Environmentally friendly" generation used by customers who want to reduce pollution
- Generation to be used to improve reliability or power quality when operational needs exceed the level of service that SCE can provide. Note: Self-generation does not include "merchant generation" intended for sale in California's wholesale electricity market. See **Merchant Generators** below.

If You Choose to Generate Your Own Power

Customers connecting generating systems, using *non-certified* or *non-pre-approved* generating equipment for SCE grid connection, must follow the Application and Interconnection Procedures outlined in **Rule 21-Generating Facility Interconnections (PDF)**.

For customers connecting **small solar or wind**-powered generating systems using *pre-certified* or *pre-approved* generating equipment for SCE grid connection, may now complete the Net Metering Application Process using our new convenient **On-Line Net Metering Application** process below. Here are frequently asked questions about **SCE's Net Energy Metering Program**.

Here is a review of the **Net Energy Metering application and interconnection agreement (PDF)**.

Safety Note: Before connecting a generating systems to the electric grid, you must receive SCE's authorization to interconnect, and your generating installation must meet all local, state, and federal codes and regulations.

Net Metering Application Options

- Solar NEM (under 10kW) Application (revised 11/09/2009)
- Solar NEM (10kW or above) Application (PDF)
- Wind NEM (under 10kW) Application (revised 11/09/2009)
- Wind NEM (10kW or above) Application (PDF)
- Generating Facility Interconnection Application (10kW or above) (PDF)

For additional information regarding Customer Generation, please contact **customer.generation@sce.com**.

You may also visit these links for more information:

- Consumer Energy Center (eligible equipment and ratings)
- California Solar Initiative (handbook)
- CSI Handbook and Forms

SCE's Self Generation Incentive Program (SGIP)

If your business is an SCE customer with a demand of 30 kilowatts (kW) or more, you can receive a cash incentive from 60 cents to $4.50 per watt for installing your own, qualifying electricity generating equipment under **SCE's Self Generation Incentive Program (SGIP)**.

The California Energy Commission's **Renewable Energy Buydown Program** provides similar incentives for renewable self-generation units under 30 kW.

*Here, you will find important new information about changes in the rules and regulations for the **SGIP program**.*

Merchant Generators

Self-generation does not include "merchant generation" intended for sale in California's wholesale electricity market. To connect a merchant generating facility to SCE's electric grid, please review the **Transmission Owner Tariff (PDF)** for interconnection at 220 kilovolts (kV) or above, or the **Wholesale Distribution Access (PDF)** Tariff for lower voltages. For additional information regarding the interconnection of your generation facility to the transmission grid, please send your questions in an email with the word "Question" typed into the subject line to **InterconnectionQA@sce.com**. We will make every effort to promptly reply to your email with a phone call or an email.

Richard L. Itteilag

Easy Energy-Saving Tips

Don't forget the basics. This simple stuff will save energy-and money-right now.

1. **Unplug**

    o   Unplug seldom-used appliances, like an extra refrigerator in the basement or garage that contains just a few items. You may save around $10 every month on your utility bill.
    o   Unplug your chargers when you're not charging. Every house is full of little plastic power supplies to charge cell phones, PDA's, digital cameras, cordless tools and other personal gadgets. Keep them unplugged until you need them.
    o   Use power strips to switch off televisions, home theater equipment, and stereos when you're not using them. Even when you think these products are off, together, their "standby" consumption can be equivalent to that of a 75 or 100 watt light bulb running continuously.

2. **Set Computers to Sleep and Hibernate**

    o   Enable the "sleep mode" feature on your computer, allowing it to use less power during periods of inactivity. In Windows, the power management settings are found on your control panel. Mac users, look for energy saving settings under system preferences in the apple menu.
    o   Configure your computer to "hibernate" automatically after 30 minutes or so of inactivity. The "hibernate mode" turns the computer off in a way that doesn't require you to reload everything when you switch it back on. Allowing your computer to hibernate saves energy and is more time-efficient than shutting down and restarting your computer from scratch. When you're done for the day, shut down.

3. **Take Control of Temperature**

    o   Set your thermostat in winter to 68 degrees or less during the daytime, and 55 degrees before going to sleep (or when you're away for the day). During the summer, set thermostats to 78 degrees or more. (for a more detailed summer energy-saving tip.)
    o   Use sunlight wisely. During the heating season, leave shades and blinds open on sunny days, but close them at night to reduce the amount of heat lost through windows. Close shades and blinds during the summer or when the air conditioner is in use or will be in use later in the day.
    o   Set the thermostat on your water heater between 120 and 130 degrees. Lower temperatures can save more energy, but you might run out of hot water or end up using extra electricity to boost the hot water temperature in your dishwasher.

4. **Use Appliances Efficiently**

    o   Set your refrigerator temperature at 38 to 42 degrees Fahrenheit; your freezer should be set between 0 and 5 degrees Fahrenheit. Use the power-save switch if your fridge has one, and make sure the door seals tightly. You can check this by making sure that a dollar bill closed in between the door gaskets is difficult to pull out. If it slides easily between the gaskets, replace them.
    o   Don't preheat or "peek" inside the oven more than necessary. Check the seal on the oven door, and use a microwave oven for cooking or reheating small items.
    o   Wash only full loads in your dishwasher, using short cycles for all but the dirtiest dishes. This saves water and the energy used to pump and heat it. Air-drying, if you have the time, can also reduce energy use.
    o   In your clothes washer, set the appropriate water level for the size of the load; wash in cold water when practical, and always rinse in cold.

- Clean the lint filter in the dryer after each use. Dry heavy and light fabrics separately and don't add wet items to a load that's already partly dry. If available, use the moisture sensor setting. (A clothesline is the most energy-efficient clothes dryer of all!)

5. **Turn Out the Lights**

    - Don't forget to flick the switch when you leave a room.
    - Remember this at the office, too. Turn out or dim the lights in unused conference rooms, and when you step out for lunch. Work by daylight when possible. A typical commercial building uses more energy for lighting than anything else.

6. **Install thermal and moisture protection systems**

    More than $400 in energy costs could be saved during the first year after home retrofits are installed. For more than 30 years, Ontario, Canada-based BSG building science experts have been educating customers on the importance of thermal and moisture protection systems as critical elements to ensure the energy efficiency and longevity of their home or commercial structure.

Emerging Energy-Efficient Industrial Technologies

N. Martin, E. Worrell, M. Ruth, L. Price (LBNL)
R. N. Elliott, A. M. Shipley, J. Thorne (ACEEE)
October 2000

Executive Summary

U.S. industry consumes approximately 37 percent of the nation's energy to produce 24 percent of the nation's GDP. Increasingly, industry is confronted with the challenge of moving toward a cleaner, more sustainable path of production and consumption, while increasing global competitiveness. Technology will be essential for meeting these challenges. At some point, businesses are faced with investment in new capital stock. At this decision point, new and emerging technologies compete for capital investment alongside more established or mature technologies. Understanding the dynamics of the decision-making process is important to perceive what drives technology change and the overall effect on industrial energy use.

The assessment of emerging energy-efficient industrial technologies can be useful for:

- identifying R&D projects;
- identifying potential technologies for market transformation activities;
- providing common information on technologies to a broad audience of policy-makers; and
- offering new insights into technology development and energy efficiency potentials.

With the support of PG&E Co., NYSERDA, DOE, EPA, NEEA, and the Iowa Energy Center, staff from LBNL and ACEEE produced this assessment of emerging energy-efficient industrial technologies. The goal was to collect information on a broad array of potentially significant emerging energy-efficient industrial technologies and carefully characterize a sub-group of approximately 50 key technologies. Our use of the term "emerging" denotes technologies that are both pre-commercial but near commercialization, and technologies that have already entered the market but have less than 5 percent of current market share. We also have chosen technologies that are energy-efficient (i.e., use less energy than existing technologies and practices to produce the same product), and may have additional "non-energy benefits." These benefits are as important (if not more important in many cases) in influencing the decision on whether to adopt an emerging technology.

The technologies were characterized with respect to energy efficiency, economics, and environmental performance. The results demonstrate that the United States is not running out of technologies to improve energy efficiency and economic and environmental performance, and will not run out in the future. We show that many of the technologies have important non-energy benefits, ranging from reduced environmental impact to improved productivity and worker safety, and reduced capital costs.

Methodology

The assessment began with the identification of approximately 175 emerging energy-efficient industrial technologies through a review of the literature, international R&D programs, databases, and studies. The review was not limited to U.S. experiences, but rather we aimed to produce an inventory of international technology developments. We devised an initial screening process to select the most attractive technologies that had: (1) high potential energy savings; (2) lower comparative first costs relative to existing technologies; and (3) other significant benefits. While some technologies scored high on all of these characteristics, most had a mixed score. We formalized this approach in a very simple rating system. Based on the literature review and the application of initial screening criteria, we identified and developed profiles for 54 technologies. The technologies ranged from highly specific ones that can be applied in a single industry to more broadly crosscutting ones that can be used in many industrial sectors.

Each of the selected technologies has been assessed with respect to energy efficiency characteristics, likely energy savings by 2015, economics, and environmental performance, as well as what's needed to further the development or implementation of the technology. The technology characterization includes a one to two-page description and a one-page table summarizing the results for the technology.

Summary of Results

Table ES-1 provides an overview of the 54 emerging energy-efficient industrial technologies. We evaluated energy savings in two ways. The third column of Table ES-1 (Total Energy Savings) shows the amount of total manufacturing energy that the technology is likely to save in 2015 in a business-as-usual scenario. The fourth column (Sector Savings) reflects the savings relative to expected energy use in the particular sector. We believe that both metrics are useful in evaluating the relative savings potential of various technologies.

Economic evaluation of the technology is identified in the summary table by simple payback period, defined as the initial investment costs divided by the value of energy savings less any changes in operations and maintenance costs. We chose this measure since it is frequently used as a shorthand evaluation metric among industrial energy managers. Payback times for the technologies range from the immediate to 20 years or more. Of the 54 technologies profiled, 31 have estimated paybacks of 3 years or less, with six paying back immediately.

Energy savings are most often not the determining factor in the decision to develop or invest in an emerging technology. Over two-thirds of technologies not only save energy but yield non-energy benefits. We separated these non-energy benefits into environmental and other categories. We assessed how important the environmental benefits are to the technology adoption decision and listed the nature of the other benefit(s). We include an assessment of the importance of these non-energy benefits.

Technologies do not seamlessly enter existing markets immediately after development. The acceptance of emerging technologies is often a slow process that entails active research and development, prototype development, market demonstration, and other activities. In Table ES-1 we summarize the recommendations for the primary activities that could be undertaken to increase the technologies' rate of uptake. Over half of these technologies have already been developed to prototype stage or are already commercial but require further demonstration and dissemination.

Each technology is at a different point in the development or commercialization process. Some technologies still need further R&D to address cost or performance issues, some are ready for demonstration, and others have already proven themselves in the field and the market needs to be informed of the benefits and market channels needed to develop skills to deliver the technology. Our outlining of recommended actions in Table ES-1 is not an endorsement of any particular technology. Technology purchasers and users will ultimately decide regarding future development. However, the actions specified are intended to help identify whether a technology is both technically and economically viable and whether it is robust enough to accommodate the stringent product quality demands in various manufacturing establishments.

Seventeen emerging technologies could benefit from additional R&D. We suggest further R&D for several primary metal technologies, and several cross-cutting motor and utility technologies. In addition to private research funds, several of the identified technologies have received some R&D support from DOE or other public entities, including federal and state agencies.

There are also a large number of technologies that already have made some headway into the marketplace or are at the prototype testing stage, and therefore are candidates for demonstration for potential customers to gain comfort with the technology. While we recommend further demonstration and dissemination of these technologies, it was often difficult to understand what is limiting their uptake without more comprehensive investigation of market issues. Some of the technologies in this category are common in European countries or Japan but have not yet penetrated the U.S. market. Others are being newly developed in the United States and face challenges in reducing the risks perceived by potential purchasers. Two technologies, motor system optimization and pump efficiency improvement, are opportunities for training programs similar to those developed by DOE for the compressed air system management. For advanced industrial CHP turbine systems, the major recommended activity is removal of policy barriers. For other technologies, their unique markets will dictate the form of the educational and promotional activities. We urge the reader to follow up on any details in the specific technology profiles provided in Section VI of this report.

We assess the technology's likelihood of success in the marketplace. While our study evaluates each technology in relation to a given reference technology, the reality of the market is that technologies compete for market share. We made a judgement (based on the energy savings, cost-effectiveness, importance of non-energy benefits, market conditions, data reliability, and potential competing technologies) as to the likelihood that the technology would succeed in the marketplace.

From a national energy policy perspective, it is important to understand which technologies have both a high likelyhood of success and a high energy-savings. While various audiences may be interested in sector-specific or regional-specific technologies, the technologies listed in Table ES-2 are intended to provide guidance to those interested in the impact of energy-saving technologies on a more national level. This table also identifies the recommended next steps appropriate for each technology.

Conclusions and Recommendations for Future Work

For this study, we identified about 175 emerging energy-efficient technologies in industry, of which we characterized 54 in detail. While many profiles of individual emerging technologies are available, few reports have attempted to impose a standardized approach to the evaluation of the technologies. This study provides a way to review technologies in an independent manner, based on information on energy savings, economic, non-energy benefits, major market barriers, likelihood of success, and suggested next steps to accelerate deployment of each of the analyzed technologies.

There are many interesting lessons to be learned from further investigation of technologies identified in our preliminary screening analysis. The detailed assessments of the 54 technologies are useful to evaluate claims made by developers, as well as to evaluate market potentials for the United States or specific regions. In this report we show that many new technologies are ready to enter the market place, or are currently under development, demonstrating that the United States is not running out of technologies to improve energy efficiency and economic and environmental performance, and will not run out in the future. The study shows that many of the technologies have important non-energy benefits, ranging from reduced environmental impact to improved productivity. Several technologies have reduced capital costs compared to the current technology used by those industries. Non-energy benefits such as these are frequently a motivating factor in bringing technologies such as these to market.

Further evaluation of the profiled technologies is still needed. In particular, further quantifying the non-energy benefits based on the experience from technology users in the field is important. Interactive effects and intertechnology competition have not been accounted for and ideally should be included in any type of integrated technology scenario, for it may help to better evaluate market opportunities.

While this report focuses on the United States, state- or region-specific analysis of technologies may provide further insights into opportunities specific for the region served. Regional specificity is determined by the type of users (i.e., industrial activities) in the region, as well as the available technology developers. Combining region-specific circumstances with technology evaluations provided in this report may lead to recognition of varying needs and the appropriate policy choices for regional (e.g., state or utility) agencies.

Our selection of a limited set of 54 technologies was an arbitrary constraint based on the funding available. A number of the initial technologies screened appeared very interesting and warrant further study, but were eliminated due to resource constraints. In addition, the initial list of candidate technologies should not be viewed as all-encompassing. The authors are aware that other promising existing technologies exist, and that by their nature new technologies will be continually emerging. Ideally, the effort reflected in this report should be the start of a continuing process that identifies and profiles the most promising emerging energy-efficient industrial technologies and tracks the market success for these technologies. An interactive database may be a better choice for it would allow the continuous updating of information, rather than providing a static snapshot of the industrial technology universe.

Richard L. Itteilag

Efficient Appliances Save Energy—and Money

Consumers get lower utility bills, and we all get a cleaner environment.

The major appliances in your home—refrigerators, clothes washers, dishwashers—account for a big chunk of your monthly utility bill. And if your refrigerator or washing machine is more than a decade old, you're spending a lot more on energy than you need to.

Today's major appliances don't hog energy the way older models do because they must meet minimum federal energy efficiency standards. These standards have been tightened over the years, so any new appliance you buy today has to use less energy than the model you're replacing. For instance, if you buy one of today's most energy-efficient refrigerators, it will use less than half the energy of a model that's 12 years old or older.

Of course, efficient appliances don't just save you money; they're good for the environment. The less energy we all use, the lower our demand on power plants, which means less pollution. The trick is to figure out which models use the least energy. Here are some guidelines.

Look for the Energy Star® label. Energy Star models are the most energy efficient in any product category, exceeding the energy efficiency minimums set by the federal government. If you remember only one rule when you shop, remember to look for the Energy Star label. In some parts of the country, utilities and state governments even sweeten the deal by offering rebates on Energy Star-rated models. Check http://www.energystar.gov for details.

Use the EnergyGuide label. Some uninformed salespeople might tell you that a model you're looking at is the most efficient because it has an EnergyGuide label. Not exactly. All new appliances must carry the EnergyGuide label, either on the appliance itself or on the packaging. The label allows you to compare the typical annual energy consumption and operating cost of different models of any type of appliance you're thinking of buying.

Get the right size. Make sure the product you're buying suits your needs. Oversized air conditioners, water heaters and refrigerators waste energy and money; in many cases they also don't perform as well.

Whenever possible choose appliances that run on natural gas rather than electricity. Usually it's more efficient to burn natural gas where it's needed—in your home—than to burn it at a power plant, convert the heat to electricity and then send the electricity over wires to your house. Look for dryers, stoves and water heaters that run on natural gas.

Think long term. Many of the most energy-efficient appliances cost more initially, but they'll save you money in the long run. Expect to keep most major appliances between 10 and 20 years. A more efficient appliance soon pays for itself; lower monthly utility bills over the lifetime of the appliance will more than offset a higher purchase price. In addition, the latest resource-efficient clothes washers and dishwashers not only save electricity, they also use a lot less water and can reduce your water bill.

Below is more specific information to keep in mind if you're in the market for any of the following major appliances.

NRDC: Setting the Standard

Energy efficiency standards may not be as high profile as saving endangered species or cleaning up toxic waste, but they are a hugely important cause for environmentalists. Since their inception, these standards have saved consumers over $200 billion—about $2,000 per household—while cutting electricity use 5 percent and reducing levels of pollution that come from the power plants that produce the electricity by over 2 percent. These savings are projected to more than double over the next 20 years even without new action. If NRDC's recommendations for new and updated standards are adopted, these savings will more than triple.

NRDC's energy program has played an important role in creating the framework under which continued improvements in appliance energy efficiency have occurred. NRDC led the negotiations that crafted the National Appliance Energy Conservation Act (1987), the law that impelled manufacturers to develop today's energy-efficient appliances.

In the early 1990s, David Goldstein, co-director of NRDC's energy program, proposed the Super Efficiency Refrigerator Program, which spurred development of the new refrigerator technology from which consumers are benefiting today. Similar programs are offered by the Consortium for Energy Efficiency. David was awarded a 2002 MacArthur Fellowship for his innovative work proving that energy efficiency makes good economic sense.

REFRIGERATORS

If you are thinking of replacing an old appliance, the refrigerator is a good place to start. New refrigerators consume 75 percent less energy than those produced in the late 1970s. A family replacing a 1980 vintage fridge with one that meets today's standards will save more than $100 a year in utility costs. Go one step further and buy an Energy Star-qualified model, and your new refrigerator will save you an additional 15 percent or more by employing better insulation, more efficient compressors and more precise temperature control and defrost mechanisms.

Energy-Saving Purchasing Tips:

- Refrigerators with freezers on top use 10 to 15 percent less energy than a side-by-side model of equivalent size.
- Generally, the larger the refrigerator, the greater the energy consumption. But one large refrigerator will use less energy than two smaller ones with the same total volume or a smaller fridge plus a separate freezer.

CLOTHES WASHERS

The energy efficiency of standard top-loading washers has doubled over the last two decades, mostly by decreasing the amount of water used. (Most of a washer's energy consumption goes to heating water.) Front-loading washers have also become more readily available. They generally use less water than top-loaders because they don't have to totally submerge clothes. Their tumbling action constantly lifts water and drops it back down onto clothing. Energy Star top-loaders, however, can be just as efficient as front loaders. Look for the EnergyGuide or Energy Star labels to compare efficiencies.

Replacing a pre-1994 washer with an Energy Star model can save a family $110 a year on utility bills. Energy Star washers use 50 percent less energy than other standard models, and only 18 to 25 gallons of water for a full-sized load, compared to 40 gallons for standard full-size washers. Many Energy Star models also advertise lower fabric wear, better stain removal and briefer drying times.

Energy-Saving Purchasing Tips:

- Choose the right size washer. A smaller washer may be more efficient for small households. But if you have a large family and have to do multiple loads in a washer that's too small for your needs, you could lose any possible energy savings.
- Look for a washer with adjustable water levels. This gives you the option of using less water to wash small loads.
- Choose a washer with a faster spin speed. This allows more water to be removed after the wash, reducing the drying time and your dryer's energy use.
- Use a gas dryer rather than an electric dryer where possible.

DISHWASHERS

A new dishwasher is not only more efficient than older models, but it's also better at getting dishes clean. Manufacturers no longer recommend that you pre-wash your dishes. Simply scrape the remaining food off your plates and place them in the machine as is. This will save you time and save money on your water bill.

The most efficient dishwashers use less hot water, have energy-efficient motors and use sensors to determine the length of the wash cycle and the water temperature needed to do the job. The newest Energy Star dishwashers are 25 percent more efficient than the minimum federal standards. Replacing a pre-1994 dishwasher with an Energy Star model can save $25 a year on utility costs.

Energy-Saving Purchasing Tips:

- Choose a dishwasher with a "light wash" or "energy-saving" wash cycle. It uses less water and operates for a shorter period of time for dishes that are just slightly soiled.
- Look for dishwashers that have an energy-saving cycle that allows dishes to be air-dried with circulation fans, rather than heat-dried with energy-wasting heating coils.

ROOM AIR CONDITIONERS

The most efficient room air conditioners have higher-efficiency compressors, fan motors and heat-transfer surfaces than previous models. A high-efficiency unit reduces energy consumption by 20 to 50 percent. Replacing a 10-year-old model with an Energy Star model can cut energy bills by an average of $14 a year.

Energy-Saving Purchasing Tips:

- Remember, the biggest unit isn't always the best choice, especially for small areas. A smaller unit running for a long period of time operates more efficiently and is more effective at decreasing humidity than a larger unit that goes on and off frequently.
- If you're comparing several similar units, choose the one with the highest Energy Efficiency Ratio. You can find the EER on the unit or its packaging. The minimum EER required by

federal law is 9.7; the most efficient air conditioners of 2003 have an EER of 11.7.

CENTRAL AIR CONDITIONERS

If your central air conditioning system is more than 10 years old, replacing it with an Energy Star model could reduce your energy consumption for cooling by 20 percent.

Energy-Saving Purchasing Tips:

- Look for the seasonal energy efficiency ratio (SEER). Old units typically have a SEER of 6 or 7. In 2006, new standards go into effect, raising the minimum SEER for central air conditioners to 13. Energy Star models already meet the SEER 13 standard, and also perform more efficiently when it's hot.
- For maximum efficiency on the hottest days, the air conditioner should have a thermal expansion valve (TVX), and the high temperature rating (EER) on your unit should be at least 11.6.
- For optimal performance, buy a matched system of indoor unit, condenser and even thermostat.
- Get a reliable contractor to make sure your new unit is the right size for your home, and have it professionally installed. Even the most efficient system can't make up for the energy loss due to improper sizing and poor installation.
- Have your contractor make sure all your ducts are sealed and insulated. Duct tests require a fan and a pressure gauge—they cannot be done by sight.

WATER HEATERS

Water heating is typically the third largest energy expense in your home, accounting for about 14 percent of your energy bill. An old water heater can operate for years at very low efficiency before it finally fails. If your gas water heater is more than 10 years old, it probably operates at less than 50 percent efficiency.

Energy-Saving Purchasing Tips:

- Calculate how much hot water your household uses at peak times. Figure that a clothes washer on hot wash/hot rinse can use about 32 gallons of hot water; a shower, 20 gallons. Washing dishes by hand can use 10 to 15 gallons, and automatic dishwashers, about 8 gallons.
- Match this figure with the "first hour rating" (FHR) on the EnergyGuide label. The FHR measures how many gallons of hot water your heater can deliver during a busy hour. Don't be misled by the size of the tank—it doesn't necessarily correlate with FHR.
- Once you've found the right FHR range for your household, check the unit's Energy Factor (EF), which rates efficiency. A high-efficiency gas model would have an EF around 0.8.
- A natural gas unit will cost less to operate than electric.

HOME ELECTRONICS

For most products, the Energy Star label is your assurance that the product will operate more efficiently than a standard model. But Energy Star TVs, audio equipment, telephones, computers and printers earn the label primarily because they draw only a small amount of power when not in use—regardless of the amount of power they consume when operating. When buying electronics, do look for the Energy Star label, but also keep a few general caveats in mind.

Energy-Saving Purchasing Tips:

- Ink jet printers tend to be more energy-efficient than lasers.
- LCD televisions and monitors draw less power than CRT or plasma screens.
- Small lightweight power supplies tend to be more energy efficient than large, heavy transformer-based power supplies.

MORE SMART SHOPPING TIPS

- **Check for incentives.** Some states offer rewards for buying the most energy-efficient appliances. Connecticut and California, for example, have rebate programs that will refund part of the purchase price of certain new energy-efficient appliances. Maryland eliminates sales tax on some appliances with the Energy Star label. Check with your local utility and the Energy Star Rebate Locator to find out if cash rebates or other incentives are available in your area, or see our state-by-state listing.
- **Use the Internet.** Several websites contain additional useful information. The EPA's Energy Star website has information on appliance models that carry the Energy Star label and where you can buy them. The American Council for an Energy Efficient Economy publishes a yearly list of the most energy-efficient appliances. And the Consortium for Energy Efficiency has information on programs promoting energy efficiency in the home.

It's a matter of supply and demand—energy prices go up as our energy use goes up, and higher prices affect "the bottom line" of every person and family.

- In 2006, each U.S. household, on average, will spend nearly $5,000 to power its home and vehicles. That's a huge chunk of any budget—and a 32 percent increase over the past two years!
- But with energy efficiency, you have the power to control your energy use without sacrificing comfort or convenience—and "insulate" yourself against future price spikes. With energy efficiency and smart energy practices, individuals and families can reduce their home energy bills and make fewer of those costly trips to the gas pump.

2. Energy use affects our home comfort—or discomfort if we don't do it right.

There are many steps you can take to reduce your energy use and costs without sacrifice or deprivation, while keeping your home comfortable.

- Save up to 20 percent on your home energy bill, according to the U.S. Environmental Protection Agency, by insulating adequately and sealing air leaks. You'll also enjoy increased indoor comfort and avoid "heating the outdoors."
- Eliminate additional drafts and cold spots by installing ENERGY STAR-labeled energy-efficient windows with double or triple panes. Depending on the climate, households can save $110 to $540 annually compared with single-pane windows. Or use storm windows.
- Using a programmable thermostat that "remembers for you" to adjust the indoor temperature according to your daily routine ensures waking up to and coming home to a comfortable house—without wasting energy and money by running the heat at full tilt overnight or while the house is empty. For every degree you lower your thermostat in winter, you can save up to 5 percent on the heating portion of your home energy bill (depending on your climate region).
- Use insulating foam sealant to seal the gaps and cracks through which cooled and heated air escape to help make your home more comfortable and energy efficient.

3. Energy use affects the air we breathe and our respiratory health.

More frequent visits to hospital emergency rooms for asthma and other respiratory conditions occur on days when ozone concentrations are high, according to the U.S. Environmental Protection Agency (EPA), which adds that in controlled studies, ozone worsened airway inflammation and caused other results likely to indicate worsening asthma.

- The prevalence of asthma in the United States has doubled in the last 20 years, according to the EPA, most rapidly in children younger than 17. More than 20 million people now report having the disease. Each year, illness associated with asthma accounts for an estimated 10 million patient visits, an estimated loss of 3 million work days, and 90 million days of restricted activity.
- Whether fueled by coal or natural gas, power plants release carbon dioxide (CO_2), noxious gases, and/or sludge as waste products as they generate electricity to power your home. Fossil fuels take millions of years to make, and their supplies are finite.
- The Alliance to Save Energy estimates that purchasing the most fuel-efficient vehicle in a particular class can save consumers $300 to $700 in annual fuel costs and considerably reduce unhealthful emissions.
- Refrigerators in the U.S. use energy equivalent to the output of some 60 300-megawatt power plants. Yet if all of the nation's households used the most efficient refrigerators, electricity savings would eliminate the need for 20 to 30 power plants—and reduce unhealthful pollution considerably.

4. **Energy use affects our nation's economic well-being, too.**

- We can help address energy-related problems by utilizing energy efficiency, which has reduced our nation's energy use by 47 percent in the past 30 years and is our greatest energy resource.
- Using less energy is also a way to generate good jobs for our economy. In its 2005 Annual Report, ENERGY STAR estimated that every federal dollar invested in its partnership programs had generated more than $60 for the US economy and created more that $15 in private sector investments. And when the National Research Council examined 17 randomly selected DOE energy efficiency R&D programs, it found $20 in economic benefits were generated by each R&D dollar expended.

- When we cut our energy use and costs, it frees up money for more productive purposes. That's as true for the nation as a whole as for an individual family. And, working together for the common good, large numbers of people taking steps to save energy—even small steps—leads to significant benefits for all of us.
- The Environmental Protection Agency estimates that if every U.S. household replaced just one traditional light with an ENERGY STAR-qualified bulb, we would save enough energy to light 7 million homes, and save $600 million in utility bills.

5. **Energy use affects our energy security.**

- The wise and efficient use of energy—no matter its source—extends our nation's energy supplies. And, when natural disasters disrupt energy supplies, being energy efficient helps us rebound quicker, better and more economically.
- As we continue to pursue energy efficiency here in the U.S., we will serve as an economic model to developing countries that are expected to require more and more energy resources in the future. By sharing our best practices, consumers in these countries will be able to use energy more wisely and won't require as much energy from the global marketplace.
- The U.S. is pioneering research and development in energy efficiency technologies that will assist energy consumers around the world and help our nation's energy security by easing the balance between global energy supply and demand.
- The U.S. sits on only 3 percent of the world's known oil reserves and accounts for 5 percent of the world's population—yet we consume 25 percent of the world's oil.
- We not only increase our national energy security when we reduce our use of energy, we also lower our own costs, improve our nation's productivity, and help curb greenhouse gas emissions.

6. **The world we leave behind.**

   - We make trade-offs for energy.

Virtually every source of energy triggers a trade-off, whether it's:

- Increased land consumption and adverse land impacts,
- Added energy production demands and higher commodity prices,
- Adverse effects on fish and wildlife,
- Waste generation,
- The air we breathe,
- Greenhouse gas emissions, or
- Visual blight.

Energy production and use account for nearly 88 percent of greenhouse gas emissions, and more environmental damage than any other human activity, according to the Alliance to Save Energy.

There is a growing recognition of the reality of global warming and the harmful effects of climate change on both our environment and public health. Ocean waters have warmed one degree Fahrenheit since 1970, and category 4 and 5 hurricanes have doubled worldwide since that decade. One recent study suggests the rate of Greenland's ice melt has more than doubled in the last decade.

We can help to address these problems by utilizing energy efficiency, which over the last 30 years has lowered our energy demand by 47 percent compared to what it would otherwise have been. Energy efficiency truly is our greatest energy resource.
Using less energy by employing energy-efficient technologies—compact fluorescent light bulbs, programmable thermostats—and engaging in smart energy practices—turning off all lights and electronics that are not in use—helps our world by reducing pollution, and also saves money for individuals, families, communities, states, and countries.

7. Bringing the *6 ° of Energy Efficiency* concept full circle, we see that once again, the global and the personal are closely related, and we can each make a difference—for ourselves and for the planet.

Chiller Program

Cool Down your Costs and Heat Up Your Savings

We know it is challenging to keep energy costs as low as possible while maintaining a comfortable temperature for employees, patrons and equipment. With FPL's Chiller Program, you can do both.

Here's how our Chiller Program works:

- We will analyze your existing chiller's efficiency to determine if you will benefit from a new high-efficiency chiller
- We will determine how much you'll save by installing a chiller in a new construction project or current facility. An energy-saving model will result in:
    o Lower ongoing operation and maintenance costs
    o Increased energy savings year after year
    o Long-term reliability
- Your FPL Customer Manager will help you through the entire process from analysis to installation

Qualifying for Incentives

- The incentive is based on chiller type, capacity, and efficiency.
- Our chiller incentives apply to qualifying high-efficiency models that have an ARI-certified rating.
- Incentive amounts and qualifying conditions vary, depending on the type and size of the equipment you replace or install.
- Back-up or emergency chillers do not qualify for incentives.

Savings In Action

- Business: Large hotel
- Size: 500,000 square feet
- Improvement: The hotel installed three 1600 ton centrifugal chillers at .53 kW/ton efficiency which is 7 percent better than required by code.
- Incentive: Received a $19,968 FPL incentive toward the chiller replacement project cost.
- Result: Projected $60,000/year in savings
- Payback: 2.5 years on programs promoting energy efficiency in the home.

IF THE CHILLER IS...	FPL PAYS...	MINIMUM QUALIFICATIONS
Air Cooled	From $5/ton to $41/ton	10 EER
Water Cooled Under 150 Tons	From $2/ton to $30/ton	.81 kW/ton Reciprocating .76 kW/ton Screw/Scroll
Water Cooled Under 300 Tons but at Least 150 Tons	From $2/ton to $24/ton	.69 kW/ton Screw/Scroll .61 kW/ton Centrifugal
Water Cooled 300 Tons and Over	From $2/ton to $17/ton	.62 kW/ton Screw/Scroll .55 kW/ton Centrifugal

AIR-COOLED CHILLER INCENTIVES		
New chiller rating (EER)*	Incentive/ton	100-ton chiller incentive
10.4	$9.50	$950
10.6	$11.39	$1,139
10.8	$13.17	$1,317
11.0	$14.95	$1,495
11.2	$16.73	$1,673
11.4	$18.51	$1,851
11.7	$20.39	$2,039

WATER-COOLED CENTRIFUGAL CHILLER INCENTIVES		
New chiller rating (EER)*	Incentive/ton	400-ton chiller incentive
.53	$4.16	$1,664
.52	$5.05	$2,020
.51	$5.94	$2,376
.50	$6.83	$2,732
.49	$7.72	$3,088
.48	$8.61	$3,444
.47	$9.50	$3,800

* EER = Energy Efficiency Ratio

Existing Renewables Facilities Program

In order to help attain the California Renewable Portfolio Standard's (RPS) goal of 20% of retail electricity generated from renewables by 2010, the California Energy Commission has developed and currently administers renewable energy incentive programs. The goal of these programs is to establish a competitive, self-sustaining renewable energy supply for California while increasing the near-term quantity of renewable energy generated in-state. The Existing Renewable Facilities Program (ERFP) is one of several program elements within the Energy Commission's Renewable Energy Program.

The purpose of the Existing Renewable Facilities Program (ERFP) is to allocate state funds to increase the competitiveness of existing (operational on or prior to September 26, 1996) in-state renewable generating facilities. For the purpose of the ERFP, self-sustainability refers to the ability of these facilities to continue operation without public funding by no later than December 31, 2011. The ERFP aims also to secure the environmental, economic and reliability benefits these facilities provide.

From 1998 to 2006, the Energy Commission set target prices and production incentive caps based on the facility's technology. Pursuant to new legislation (Senate Bill 1250 [Perata, Chapter 512, Statues of 2006] PDF file), the Energy Commission is now required to evaluate and consider each existing renewable facility on an individual case-by-case basis. Facilities receive funding based on production incentives (cent(s) per kilowatt hour). A target price and incentive cap are assigned to each facility based on need. If the market price for energy of a facility drops below the target price, then the Energy Commission will incentivize the facility for each kilowatt hour generated up to a maximum incentive cap.

A hypothetical example can illustrate how incentives are paid from the ERFP. For this example, assume that the Energy Commission determined that a facility has a 6-cent per kilowatt hour (kWh) target price and a 1.5 cents/kWh cap. If the market price for the facility drops to 5-cents/kWh in January, the facility would receive 1.0-cent per kWh from the ERFP for generation produced in January (5-cent/kWh market price plus one-cent/kWh incentive payment to bring the revenue stream up to the 6-cent/kWh target price). To continue the illustrative example, if the market price dropped to 4 cents/kWh in February, then the ERFP would pay 1.5 cents/kWh. Thus, in February, the facility would receive the maximum 1.5-cent per kWh cap from the ERFP plus 4-cents/kWh from the market, for a total of 5.5-cents/kWh.

ERFP eligible technologies include:

1. Solid-fuel biomass
2. Solar thermal electric
3. Wind power (due to the current market climate, these facilities do not require Energy Commission funding at this time).

This program appropriates 20% of deposited funds into the Renewable Resource Trust Fund per Senate Bill 1036 [Perata, Chapter 685, Statutes of 2007—(PDF file)]. It is estimated that approximately $75 million would be allocated to the ERFP for calendar years 2007 through 2011.

Forrests-Farms & Global Warming (Oct. 2010)

U.S. forests sequester enough carbon every year to offset roughly 11 percent of the country's industrial greenhouse gas emissions, according to a new federal report. Shifting land-use patterns, particularly the 20th century abandonment of farms in the East and the subsequent regrowth of woodlands, have helped turn U.S. forests as a whole into a carbon sink, meaning they absorb more carbon dioxide from the atmosphere than they release through natural processes. From a global warming perspective, that is a welcome trend and one that creates the potential for forest owners to sell credits on emerging carbon markets. But it also raises questions of how, and whether, forests can be managed to maintain that role. The majority of the nation's 797 million acres of forestland is in private ownership, much if it in the East, where development is chewing into the woods. "There is a real concern that the continued loss of open space and development into forest lands in the East, in particular, is going to reduce the amount of carbon we have in our forests," said U.S. Agriculture Deputy Under Secretary Ann Bartuska. On Western public lands, wildfire and tree mortality from bark beetle epidemics could put a dent in forest carbon storage. Before settlement, woodlands were probably in carbon equilibrium, releasing about as much carbon as they took up, said David Cleaves, U.S. Forest Service climate change adviser. When the settlers cleared vast expanses of forest and burned trees as fuel, a long-term rise in forest emissions was set in motion, peaking in the early 1900s. That started to reverse as the woods reclaimed fields and a federal policy of suppressing wildfire allowed dense, young stands to grow, increasing not only the amount of woodland, but also the volume of carbon stored per acre. Now the nation's forests bank an estimated 41.4 billion metric tons of carbon, roughly the equivalent of 20 years of U.S. fossil fuel emissions, Cleaves said. Every year, the Forest Service says new growth absorbs carbon dioxide in quantities that are the equivalent of taking 135 million cars off the road. The report can be found at http://www.fs.fed.us/rmrs/forest-carbon/.

Veteran west-side farmer John Diener has always felt confident in his ability to grow quality tomatoes, almonds and wheat—but to some, that may not be good enough.

Responding to consumer sentiments, grocery-chain buyers are pushing Diener and other farmers to show they practice "sustainable" agriculture—a popular if still fuzzy concept.

While similar to organic farming, its focus is broader: In contrast to conventional farming, sustainable agriculture puts greater emphasis on practices that have long-term benefits. For example, instead of using harsh chemicals, some farmers rely on parasitic insects to battle bad bugs. Or they use renewable energy rather than fossil fuels. Others work on improving the standard of living for farmworkers, ensuring a more productive and stable labor force.

The goal of sustainability is to reduce farming's impact on the environment while ensuring a future for agriculture.

And while some may disagree with how it is defined and measured, one thing is clear: "Sustainability" is changing how farmers do business.

Walmart, the world's largest retailer, announced on Oct. 14 a global plan to train 1 million farmers and workers on crop selection and sustainable-farming practices, including using water, pesticides and fertilizer more efficiently.

It's not alone. Sysco—a global supplier of food to commercial kitchens—has a sustainable-farming program, as does Del Monte Foods. (NYSE:DLM)

"This is not an issue that is going away, and it's one that more retailers will likely adopt," said Gail Feenstra, food system coordinator with the Sustainable Agriculture Research and Education Program at UC Davis. "It is best that farmers get out ahead of the game to the extent that they can."

Feenstra said climate change and high energy costs have caused retailers to take a harder look at how they do business, including their supply chain.

"The sustainability of their own operations relies on them getting products from farmers," Feenstra said. "And that isn't going to happen if the soil and air are contaminated."

Others suggest that the drive for sustainability is also fueled by the demands of Wall Street.

"If you look at the Fortune 500 companies, they are responding to consumer pressures and global pressures for resources," said Barbara Meister, marketing manager for SureHarvest, a company that provides sustainability solutions such as software and certification. "And they are responding to their stockholders who are asking about how a company is doing in the way of treatment of workers, its carbon emissions and use of water."

Making the grade

Diener's Red Rock Ranch in Firebaugh grows certified sustainable tomatoes that he sells to processor Tomatek.

That means his farm is audited once a year by Food Alliance, a Portland-based nonprofit certification group.

Inspectors verify Diener's farming practices.

He gets points for using water-saving irrigation equipment, applying fewer pesticides and providing his 20 employees with health insurance and a retirement plan.

At the end of the inspection, he either passes or he fails.

"It is like going to school," Diener said. "And as long as I get something better than a C, I am OK."

Putting up with this kind of scrutiny has become part of doing business for Diener, whom many consider a progressive farmer.

"You could say that you don't want to go through something like this," Diener said. "But then who will you sell to? In some cases, we don't have a choice."

Advocates of sustainability say much work remains in the sustainable farming movement, including establishing a national standard like one that exists for organic food.

But creating such a standard could take years. In the meantime, retailers and industry groups are coming up with their own definitions and measurement tools.

Food Alliance is among a handful of organizations that provide verification of sustainable-farming practices as defined by university research.

The nonprofit group uses a third-party auditing firm for farm and factory inspections.

It works for more than 360 farms and ranches, covering 6.5 million acres in 24 states. It also has certified 35 food packing and processing plants.

"One of our biggest concerns is that as this starts to percolate in the industry, how do you ensure that the sustainability claims being made are credible?" said Matthew Buck, assistant director of Food Alliance. "A company can make a claim that they are eco-friendly based on a standard that may not be very clear."

Industry groups—including the Almond Board of California, the California Sustainable Winegrowing Alliance and Lodi-Woodbridge Winegrape Commission—are working with their members on sustainable-farming programs.

The produce industry also has launched the Stewardship Index for Specialty Crops, which is intended to help growers measure their sustainability performance.

The organizations help educate growers about sustainable-agriculture practices such as how to use fewer pesticides, save water and reduce farm pollution.

Overcoming obstacles

But industry leaders admit that the path to sustainability is not smooth.

Growers may be slow to jump on board if the practices add to their time and production costs, without an immediate financial return.

The Lodi-Woodbridge Winegrape Commission now has about 23,000 acres in its sustainability program, up from 15,000 acres last year.

Mark Chandler, executive director of the commission, said that figure would be higher if not for two major roadblocks: the $2,000 cost of certification, and the need for detailed paperwork.

"Some just don't want to go through the hassle of writing up a plan, and others don't like the cost," Chandler said.

So far, only two wineries in the region pay growers a bonus for being certified by a third-party auditor.

Modesto tree fruit farmer Paul Van Konynenburg isn't sure what the future of sustainable farming will look like. All he knows is that much has changed in the 20 years he has been farming apples, cherries, peaches and apricots.

"We are dealing in a whole new world now, where we have things like sustainability audits," Van Konynenburg said. "And if a Walmart or Costco (NASDAQ:COST) or Safeway (NYSE:SWY) say you have to have it, you say, 'Yes sir.'

But Van Konynenburg said that may not be such a bad thing, "because what this comes down to for me is giving the consumer something they can trust. And that is going to benefit all of us."

Heat pump

Outdoor section of a residential electric air-source heat pump is an ideal load-leveling technique for electric utilities to balance the company's generation capacity by causing the consumer to utilize electricity off-peak in the winter.

A **heat pump** is a machine or device that moves heat from one location (the 'source') at a lower temperature to another location (the 'sink' or 'heat sink') at a higher temperature using mechanical work or a high-temperature heat source. The difference between a heat pump and a normal air conditioner is that a heat pump can be used to provide heating or cooling. Even though the heat pump can heat, it still uses the same basic refrigeration cycle to do this. In other words a heat pump can change which coil is the condenser and which the evaporator. This is normally achieved by a reversing valve. In cooler climates it is common to have heat pumps that are designed only to provide heating.

Common examples are food refrigerators and freezers, air conditioners, and reversible-cycle heat pumps for providing building space heating. In heating, ventilation, and air conditioning (HVAC) applications, a heat pump normally refers to a vapor-compression refrigeration device that includes a reversing valve and optimized heat exchangers so that the direction of heat flow may be reversed. Most commonly, heat pumps draw heat from the air or from the ground.

Overview

Heat pumps have the ability to move heat energy from one environment to another, and in either direction. This allows the heat pump to both bring heat into an occupied space, and take it out. In the cooling mode a heat pump works the same as an ordinary air conditioner (A/C). It uses an evaporator to absorb heat from inside an occupied space and rejects this heat to the outside through the condenser. The refrigerant flows outside of the space to be conditioned, where the condenser and compressor are located, while the evaporator is inside. The key component that makes a heat pump different from an A/C is the reversing valve. The reversing valve allows for the flow direction of the refrigerant to be changed. This allows the heat to be pumped in either direction.

- In **heating mode** the outdoor coil becomes the evaporator, while the indoor becomes the condenser which absorbs the heat from the refrigerant and dissipates to the air flowing through it. The air outside even at 0°C has heat energy in it. With the refrigerant flowing in the opposite direction the evaporator (outdoor coil) is absorbing the heat from the air and moving it inside. Once it picks up heat it is compressed and then sent to the condenser (indoor coil). The indoor coil then rejects the heat into the air handler, which moves the heated air through out the house.

- In **cooling mode** the outdoor coil is now the condenser. This makes the indoor coil now the evaporator. The indoor coil is now the evaporator in the sense that it is going to be used to absorb the heat from inside the enclosed space. The evaporator absorbs the heat from the inside, and takes it to the condenser where it is rejected into the outside air.

Operating principles

Since the heat pump or refrigerator uses a certain amount of work to move the refrigerant, the amount of energy deposited on the hot side is greater than taken from the cold side. One common type of heat pump works by exploiting the physical properties of an evaporating and condensing fluid known as a refrigerant.

A simple stylized diagram of a heat pump's vapor-compression refrigeration cycle: 1)condenser, 2)expansion valve, 3)evaporator, 4)compressor.

The working fluid, in its gaseous state, is pressurized and circulated through the system by a compressor. On the discharge side of the compressor, the now hot and highly pressurized vapor is cooled in a heat exchanger, called a condenser, until it condenses into a high pressure, moderate temperature liquid. The condensed refrigerant then passes through a pressure-lowering device also called a metering device like an expansion valve, capillary tube, or possibly a work-extracting device such as a turbine. The low pressure, liquid refrigerant leaving the expansion device enters another heat exchanger, the evaporator, in which the fluid absorbs heat and boils. The refrigerant then returns to the compressor and the cycle is repeated.

In such a system it is essential that the refrigerant reach a sufficiently high temperature when compressed, since the second law of thermodynamics prevents heat from flowing from a cold fluid to a hot heat sink. Practically, this means the refrigerant must reach a temperature greater than the ambient around the high-temperature heat exchanger. Similarly, the fluid must reach a sufficiently low temperature when allowed to expand, or heat cannot flow from the cold region into the fluid, i.e. the fluid must be colder than the ambient around the cold-temperature heat exchanger. In particular, the pressure difference must be great enough for the fluid to condense at the hot side and still evaporate in the lower pressure region at the cold side. The greater the temperature difference, the greater the required pressure difference, and consequently the more energy needed to compress the fluid. Thus as with all heat pumps, the Coefficient of Performance (amount of heat moved per unit of input work required) decreases with increasing temperature difference.

Insulation is used to reduce the work and energy required to achieve and maintain a lower temperature in the cooled space.

Due to the variations required in temperatures and pressures, many different refrigerants are available. Refrigerators, air conditioners, and some heating systems are common applications that use this technology.

Heat sources

Many heat pumps also use an auxiliary heat source for heating mode. This means that, even though the heat pump is the primary source of heat, another form is available as a back-up. Electricity, oil, or gas are the most commons sources. This is put in place so that if the heat pump fails or can't provide enough heat, the auxiliary heat will kick on to make up the difference.

Geothermal heat pumps use the ground and water to work as the condensers and evaporators. They work in the same manner as an air to air heat pump, but instead of indoor and outdoor coils they use the earth as natural evaporators and condensers. These are very eco-friendly and are a cheaper alternative in the long run due to lower operating cost.

Applications

In HVAC applications, a heat pump normally refers to a vapor-compression refrigeration device that includes a reversing valve and optimized heat exchangers so that the direction of heat flow may be reversed. The reversing valve switches the direction of refrigerant through the cycle and therefore the heat pump may deliver either heating or cooling to a building. In the cooler climates the default setting of the reversing valve is heating. The default setting in warmer climates is cooling. Because the two heat exchangers, the condenser and evaporator, must swap functions, they are optimized to perform adequately in both modes. As such, the efficiency of a reversible heat pump is typically slightly less than two separately optimized machines.

In plumbing applications, a heat pump is sometimes used to heat or preheat water for swimming pools or domestic water heaters.

In somewhat rare applications, both the heat extraction and addition capabilities of a single heat pump can be useful, and typically results in very effective use of the input energy. For example, when an air cooling need can be matched to a water heating load, a single heat pump can serve two useful purposes. That is, a heat pump domestic water heater located in the living area of a home could cool the home, reducing or eliminating the need for additional air conditioning. This installation would be best-suited to a climate that is warm or hot most of the year. Unfortunately, these situations are rare because the demand profiles for heating and cooling are often significantly different.

Refrigerants

Until the 1990s, the refrigerants were often chlorofluorocarbons such as R-12 (dichlorodifluoromethane), one in a class of several refrigerants using the brand name Freon, a trademark of DuPont. Its manufacture was discontinued in 1995 because of the damage that CFCs cause to the ozone layer if released into the atmosphere. One widely adopted replacement refrigerant is the hydrofluorocarbon (HFC) known as R-134a (1,1,1,2-tetrafluoroethane). R-134a is not as efficient as the R-12 it replaced (in automotive applications) and therefore, more energy is required to operate systems utilizing R-134a than those using R-12. Other substances such as liquid R-717 ammonia are widely used in large-scale systems, or occasionally the less corrosive but more flammable propane or butane, can also be used.

Since 2001, carbon dioxide, R-744, has increasingly been used, utilizing the transcritical cycle. In residential and commercial applications, the hydrochlorofluorocarbon (HCFC) R-22 is still widely used, however, HFC R-410A does not deplete the ozone layer and is being used more frequently. Hydrogen, helium, nitrogen, or plain air is used in the Stirling cycle, providing the maximum number of options in environmentally friendly gases. More recent refrigerators are now exploiting the R600A which is isobutane, and does not deplete the ozone and is friendly to the environment.

Efficiency

When comparing the performance of heat pumps, it is best to avoid the word "efficiency" which has a very specific thermodynamic definition. The term coefficient of performance (COP) is used to describe the ratio of useful heat movement to work input. Most vapor-compression heat pumps utilize electrically powered motors for their work input. However, in most vehicle applications, shaft work, via their internal combustion engines, provide the needed work.

When used for heating a building on a mild day of say 10°C, a typical air-source heat pump has a COP of 3 to 4, whereas a typical electric resistance heater has a COP of 1.0. That is, one joule of electrical energy will cause a resistance heater to produce one joule of useful heat, while under ideal conditions, one joule of electrical energy can cause a heat pump to move much more than one joule of heat from a cooler place to a warmer place.

Note that when there is a wide temperature differential, e.g., when an air-source heat pump is used to heat a house on a very cold winter day of say 0°C, it takes more work to move the same amount of heat indoors than on a mild day. Ultimately, due to Carnot efficiency limits, the heat pump's performance will approach 1.0 as the outdoor-to-indoor temperature difference increases. This typically occurs around -18 °C (0 °F) outdoor temperature for air source heat pumps. Also, as the heat pump takes heat out of the air, some moisture in the outdoor air may condense and possibly freeze on the outdoor heat exchanger. The system must periodically melt this ice. In other words, when it is extremely cold outside, it is simpler, and wears the machine less, to heat using an electric-resistance heater than to strain an air-source heat pump.

Geothermal heat pumps, on the other hand, are dependent upon the temperature underground, which is "mild" (typically 10°C at a depth of more than 1.5m for the UK) all year round. Their COP is therefore is normally in the range of 4.0 to 5.0.

The design of the evaporator and condenser heat exchangers is also very important to the overall efficiency of the heat pump. The heat exchange surface areas and the corresponding temperature differential (between the refrigerant and the air stream) directly affect the operating pressures and hence the work the compressor has to do in order to provide the same heating or cooling effect. Generally the larger the heat exchanger the lower the temperature differential and the more efficient the system. Since heat exchangers are expensive, and the heat pump industry is very financially competetive, the drive towards more efficient heat pumps and air conditioners is often led by legislative measures on minimum efficiency standards.

In cooling mode a heat pump's operating performance is described as its energy efficiency ratio (EER) or seasonal energy efficiency ratio (SEER), and both measures have units of BTU/(h·W) (1 BTU/(h·W) = 0.293 W/W). A larger EER number indicates better performance. The manufacturer's literature should provide both a COP to describe performance in heating mode and an EER or SEER to describe performance in cooling mode. Actual performance varies, however, and depends on many factors such as installation, temperature differences, site elevation, and maintenance.

Heat pumps are more *effective* for heating than for cooling if the temperature difference is held equal. This is because the compressor's input energy is largely converted to useful heat when in heating mode, and is discharged along with the moved heat via the condenser. But for cooling, the condenser is normally outdoors, and the compressor's dissipated work is rejected rather than put to a useful purpose.

For the same reason, opening a food refrigerator or freezer heats up the room rather than cooling it because its refrigeration cycle rejects heat to the indoor air. This heat includes the compressor's dissipated work as well as the heat removed from the inside of the appliance.

The COP for a heat pump in a heating or cooling application, with steady-state operation, is:

where

- ΔQ_{cool} is the amount of heat extracted from a cold reservoir at temperature T_{cool},

- ΔQ_{hot} is the amount of heat delivered to a hot reservoir at temperature T_{hot}.
- ΔA is the compressor's dissipated work.
- All temperatures are absolute temperatures usually measured in <u>kelvins</u> (K).

COP and lift

The COP increases as the temperature difference, or "lift", decreases between heat source and destination. The COP can be maximised at design time by choosing a heating system requiring only a low final water temperature (e.g. underfloor heating), and by choosing a heat source with a high average temperature (e.g. the ground). Domestic hot water (DHW) and radiators require high water temperatures, affecting the choice of heat pump technology.

Pump type and source	Typical use case	COP variation with Output Temperature					
		35 °C (e.g. heated screed floor)	45 °C (e.g. heated screed floor)	55 °C (e.g. heated timber floor)	65 °C (e.g. radiator or DHW)	75 °C (e.g. radiator & DHW)	85 °C (e.g. radiator & DHW)
High efficiency air source heat pump (ASHP). Air at -20 °C		2.2	2.0	-		-	-
Two-stage ASHP air at -20 °C	Low source temp.	2.4	2.2	1.9	-	-	-
High efficiency ASHP air at 0 °C	Low output temp.	3.8	2.8	2.2	2.0	-	-
Prototype transcritical CO_2 (R744) heat pump with tripartite gas cooler, source at 0 °C	High output temp.	3.3	-	-	4.2	-	3.0

Ground source heat pump (GSHP). Water at 0 °C		5.0	3.7	2.9	2.4	-	-
GSHP ground at 10 °C	Low output temp.	7.2	5.0	3.7	2.9	2.4	-
Theoretical Carnot cycle limit, source -20 °C		5.6	4.9	4.4	4.0	3.7	3.4
Theoretical Carnot cycle limit, source 0 °C		8.8	7.1	6.0	5.2	4.6	4.2
Theoretical Lorentz cycle limit (CO_2 pump), return fluid 25 °C, source 0 °C		10.1	8.8	7.9	7.1	6.5	6.1
Theoretical Carnot cycle limit, source 10 °C		12.3	9.1	7.3	6.1	5.4	4.8

Types

The two main types of heat pumps are compression heat pumps and absorption heat pumps. Compression heat pumps always operate on mechanical energy (through electricity), while absorption heat pumps may also run on heat as an energy source (through electricity or burnable fuels). An absorption heat pump may be fueled by natural gas or LP gas, for example. While the Gas Utilization Efficiency in such a device, which is the ratio of the energy supplied to the energy consumed, may average only 1.5, that is better than a natural gas or LP gas furnace, which can only approach 1. Although an absorption heat pump may not be as efficient as an electric compression heat pump, an absorption heat pump fueled by natural gas may be advantageous in locations where electricity is relatively expensive and natural gas is relatively inexpensive. A natural gas-fired absorption heat pump might also avoid the cost of an electrical service upgrade which is sometimes necessary for an electric heat pump installation. In the case of air-to-air heat pumps, an absorption heat pump might also have an advantage in colder regions, due to a lower minimum operating temperature. ROBUR heat pumps comparison

A number of sources have been used for the heat source for heating private and communal buildings.

- air source heat pump (extracts heat from outside air)
    - air-air heat pump (transfers heat to inside air)
    - air-water heat pump (transfers heat to a tank of water)

- geothermal heat pump (extracts heat from the ground or similar sources)
    - geothermal-air heat pump (transfers heat to inside air)
        - ground-air heat pump (ground as a source of heat)
        - rock-air heat pump (rock as a source of heat)
        - water-air heat pump (body of water as a source of heat)
    - geothermal-water heat pump (transfers heat to a tank of water)
        - ground-water heat pump (ground as a source of heat)
        - rock-water heat pump (rock as a source of heat)
        - water-water heat pump (body of water as a source of heat)

Heat sources

Most commonly, heat pumps draw heat from the air (outside or inside air) or from the ground (groundwater or soil). The heat drawn from the ground is in most cases stored solar heat, and it should not be confused with geothermal heat, though the latter will contribute in some small measure to all heat in the ground. Other heat sources include water; nearby streams and other natural water bodies have been used, and sometimes domestic waste water which is often warmer than the ambient temperature.

Air-source heat pumps

Air source heat pumps are relatively easy (and inexpensive) to install and have therefore historically been the most widely used heat pump type. However, they suffer limitations due to their use of the outside air as a heat source or sink. The higher temperature differential during periods of extreme cold or heat leads to declining efficiency, as explained above. In mild weather, COP may be around 4.0, while at temperatures below around -8 °C (17 °F) an air-source heat pump can achieve a COP of 2.5 or better, which is considerably more than the COP that may be achieved by conventional heating systems. The average COP over seasonal variation is typically 2.5-2.8, with exceptional models able to exceed 6.0 (2.8 kW).

Ground source heat pumps

Ground source heat pumps, which are also referred to as Geothermal heat pumps, typically have higher efficiencies than air-source heat pumps. This is because they draw heat from the ground or groundwater which is at a relatively constant temperature all year round below a depth of about eight feet (2.5 m). This means that the temperature differential is lower, leading to higher efficiency. Ground-source heat pumps typically have COPs of 3.5-4.0 at the beginning of the heating season, with lower COPs as heat is drawn from the ground. The trade off for this improved performance is that a ground-source heat pump is more expensive to install due to the need for the digging of wells or trenches in which to place the pipes that carry the heat exchange fluid. When compared versus each other, groundwater heat pumps are generally more efficient than heat pumps using heat from the soil. Their efficiency can be further improved, by pumping summer heat into the ground. One way is to use ground water to cool the floors on hot days. Another way is to make large solar collectors, for instance by putting plastic pipes just under the roof tiles or in the tarmac of the parking lot. The most price effective way is to put a large air to water heat exchanger on the rooftop.

Solid state heat pumps

In 1881, the German physicist Emil Warburg put a block of iron into a strong magnetic field and found that it increased very slightly in temperature. Some commercial ventures to implement this technology are underway, claiming to cut energy consumption by 40% compared to current domestic refrigerators. The process works as follows: Powdered gadolinium is moved into a magnetic field, heating the material by 2 to 5 °C (4 to 9 °F). The heat is removed by a circulating fluid. The material is then moved out of the magnetic field, reducing its temperature below its starting temperature.

Solid state heat pumps using the Thermoelectric Effect have improved over time to the point where they are useful for certain refrigeration tasks. Commercially available technologies have efficiencies that are currently well below that of mechanical heat pumps, however this area of technology is currently the subject of active research in materials science.

Near-solid-state heat pumps using Thermoacoustics are commonly used in cryogenic laboratories.

History

This section requires expansion.

Milestones:
- 1748: William Cullen demonstrates artificial refrigeration.
- 1834: Jacob Perkins builds a practical refrigerator with diethyl ether.
- 1852: Lord Kelvin describes the theory underlying heat pump.
- 1855-1857: Peter Ritter von Rittinger develops and builds the first heat pump.

See also

- Ice-Stick Hybrid combining air and a groundloop
- EcoCute domestic heat pump water heater

- Flash evaporation
- Geothermal heat pump
- Heat exchanger
- Renewable heat
- Thermoelectric heat pumps that use the Peltier effect
- Vapor-compression refrigeration
- Vortex tube
- IEA-ECBCS Annex 48 : Heat Pumping and Reversible Air Conditioning

References

1. The Systems and Equipment volume of the ASHRAE Handbook, ASHRAE, Inc., Atlanta, GA, 2004
2. The Canadian Renewable Energy Network 'Commercial Earth Energy Systems', Figure 29. Retrieved December 8, 2009.
3. Technical Institute of Physics and Chemistry, Chinese Academy of Sciences 'State of the Art of Air-source Heat Pump for Cold Region', Figure 5. Retrieved April 19, 2008.
4. SINTEF Energy Research 'Integrated CO2 Heat Pump Systems for Space Heating and DHW in low-energy and passive houses', J. Steen, Table 3.1, Table 3.3. Retrieved April 19, 2008.
5. http://www2.vlaanderen.be/economie/energiesparen/doc/brochure_warmtepomp.pdf
6. Homeowners using heat pump systems[dead link]
7. "Heat pumps sources including groundwater, soil, outside and inside air)" (PDF). http://www2.vlaanderen.be/economie/energiesparen/doc/folder_warmtepomp.pdf. Retrieved 2010-06-02.
8. "Carrier web site: Heat Pumps". Residential.carrier.com. http://www.residential.carrier.com/products/acheatpumps/heatpumps/index.shtml. Retrieved 2010-06-02.
9. "the IPCC 4th Working Group III report" (PDF). http://www.ipcc.ch/pdf/assessment-report/ar4/wg3/ar4-wg3-chapter6.pdf. Retrieved 2010-06-02.

10. Guardian Unlimited, December 2006 'A cool new idea from British scientists: the magnetic fridge'
11. Banks, David L. An Introduction to Thermogeology: Ground Source Heating and Cooling. Wiley-Blackwell. ISBN 978-1-4051-7061-1.

High Cost Of Renewables (Nov. 2010)

Michael Polsky's wind farm company was doing so well in 2008 that banks were happy to lend millions for his effort to light up America with clean electricity.

But two years later, Mr. Polsky has a product he is hard-pressed to sell.

His company, Invenergy, had a contract to sell power to a utility in Virginia, but state regulators rejected the deal, citing the recession and the lower prices of natural gas and other fossil fuels.

"The ratepayers of Virginia must be protected from costs for renewable energy that are unreasonably high," the regulators said. Wind power would have increased the monthly bill of a typical residential customer by 0.2 percent.

Even as many politicians, environmentalists and consumers want renewable energy and reduced dependence on fossil fuels, a growing number of projects are being canceled or delayed because governments are unwilling to add even small amounts to consumers' electricity bills.

Deals to buy renewable power have been scuttled or slowed in states including Florida, Idaho and Kentucky as well as Virginia. By the end of the third quarter, year-to-date installations of new wind power dropped 72 percent from 2009 levels, according to the American Wind Energy Association, a trade group.

Mr. Polsky calls the focus on short-term costs short-sighted.

"They have to look for the ratepayers' long-term interest," he said, "not just the bills this year."

Electricity generated from wind or sun still generally costs more—and sometimes a lot more—than the power squeezed from coal or natural gas. Prices for fossil fuels have dropped in part because the recession has reduced demand. In the case of natural gas, newer drilling techniques have opened the possibility of vast new supplies for years to come.

The gap in price can pit regulators, who see their job as protecting consumers from unreasonable rates, against renewable energy developers and utility companies, many of which are willing to pay higher prices now to ensure a broader energy portfolio in the future.

In April, for example, the state public utilities commission in Rhode Island rejected a power-purchase deal for an offshore wind project that would have cost 24.4 cents a kilowatt-hour. The utility now pays about 9.5 cents a kilowatt hour for electricity from fossil fuels.

The state legislature responded by passing a bill allowing the regulators to consider factors other than price. The commission then approved an agreement to buy electricity from a smaller wind farm, although that decision is being challenged in the courts.

Similarly, in Kentucky this year, the public service commission voted down a contract for a local utility, Kentucky Power, to buy electricity from NextEra Energy Resources in Illinois.

According to the commission, Kentucky Power argued that the contract would position the utility "to better meet growing environmental requirements and impending government portfolio mandates for renewable energy" and that it would benefit customers.

But Kentucky's attorney general, Jack Conway, joined by business and industrial electricity users, opposed the deal, contending that it would have increased a typical residential customer's rates by about 0.7 percent and was "a discretionary expense" that the utility's customers could ill afford.

Commissioner James W. Gardner, the lone dissenting commissioner, protested that "there is a necessity for this power" and said that "there are great pressures nationally and in Kentucky to increase renewables."

Companies that make solar cells and wind machines argue that a national energy policy is needed to guarantee them a market that will allow their industry to develop. Clean power will be an important industry globally for years, they say, and if the United States does not subsidize renewable energy now, it risks falling far behind other countries.

They point to China, which is rapidly increasing the amount of electricity it generates from renewable sources. In its most recent quarterly assessment of the renewable energy sector, the accounting and consulting firm Ernst & Young identified China as the most attractive market for investment in renewable energy.

In part, the analysis suggested, this reflected the failure of American lawmakers to pass a national renewable energy standard and the looming expiration of a Treasury program that allowed renewable developers to receive cash grants in lieu of tax credits.

In Europe, many national governments have guaranteed prices for energy from sun or wind. As a result, renewable advocates say, many countries are on track to meet the European Union's goal of 20 percent of energy from renewable sources by 2020.

The United States has relied on a combination of state renewable energy mandates and federal tax credits to encourage greater reliance on energy from renewable sources. Legislation that would have set a price on carbon-dioxide emissions and included a standard for increasing the share of clean energy in the nation's electricity portfolio failed in Congress this year.

"Our investors tell us they're nervous about all the uncertainty," said John Cusack, the president of Gifford Park Associates, a sustainability management and investment consulting firm in Eastchester, N.Y. "They don't know what's going to happen."

To be sure, a lot of renewable power development is still going forward. The American Wind Energy Association estimates that wind farms capable of producing 6,300 megawatts of wind power are under construction, and that a busy second half of 2010 would leave installations about 50 percent behind last year. Solar power is becoming less expensive, and its use is expanding rapidly. But it still accounts for less than 1 percent of the nation's electricity needs, providing enough to serve about 350,000 homes.

Renewable energy supporters argue that higher fossil fuel prices will eventually make renewable energy more competitive—and at times over the last two decades, when the price for natural gas has spiked, wind power in particular has been a relative bargain. Advocates also argue that while the costs might be higher now, as the technology matures and supply chains and manufacturing bases take root, clean sources of power will become more attractive.

Fold in the higher costs of extracting and burning fossil fuels on human health, the climate and the environment, many advocates argue, and renewable technologies like wind power are already cheaper.

"One of the problems in the United States is that we haven't been willing to confront the tough questions," said Paul Gipe, who sits on the steering committee of the Alliance for Renewable Energy, a group advocating energy policy reform.

"We have to ask ourselves, 'Do we really want renewables?'" he said. "And if the answer to that is yes, then we're going to have to pay for them."

Heroes wanted in climate science story

Nov 8, 2010   USA Today
Dan Vergano

In a house in the woods, somewhere far away, perhaps lives someone who doesn't love a good story.

"Deep in our nature" lurks a love of story-telling, wrote the Greek philosopher Aristotle around 350 B.C., the world's first literary critic.

And psychologists and neuroscientists have increasingly backed up ol' Aristotle, looking at story-telling as something fundamentally human. Brain scan studies, for example, show listening to stories lights up more and different areas as children age. Alzheimer's patients loss of the ability to follow stories may be the most debilitating aspect of their dementia.

But despite the narrative neuroscience, some groups of scientists, particularly climate researchers, might want to polish their story-telling skills. Where 97% of active climate scientists agree climate change is a reality and only 52% of the public say they agree, according to an Eos journal survey, something may have gone wrong in how scientists communicate to the public.

"There's a narrative vacuum that needs to be filled," wrote the science writer Keith Kloor last month. One catch is that scientists simply prize facts over stories, as climate scientist Gavin Schmidt of NASA's Goddard Institute for Space Studies, noted last week on the "RealClimate" blog.

So is that the problem?

"Scientists, academics, and politicians on the left, do not do stories very well," says Harvard political scientist Michael Jones, who earlier this year led a Policy Studies Journal report on the use and misuse of narrative in policymaking. "You have to tell a story, though, if you want people to retain information."

Work that Jones did as a graduate student published this year, involved experiments on 1,586 people to show how this plays out in the way people talk about climate science. Each person was randomly treated to one of four opinion articles and answered survey questions about their climate opinions before and after reading the article. Each article discussed a recent report on the U.S. effects of global warming.

One of the four was simply a list of the effects of climate change from Alaska to the Atlantic Ocean, and points in between, such as "It is 66% likely that the Great Plains area will experience more severe summer droughts."

The other three options were all identically-worded stories, with the same facts as the list, but with the good guys, bad guys and solution for global warming swapped out. The options they looked at:

"Individualist" story—presented "free competition" as the hero of the story, with "bureaucratic unions" and "the infamous Club of Rome" as the enemy, with a market system as the solution to global warming.

"Hierarchical" story—presented "scientific expertise" as the hero of the story, with "Ecodefense" and the "infamous Earthfirst!" as the enemy, with nuclear energy as the solution to global warming.

"Egalitarian" story—presented "equal participation" as the hero of the story, with "the radical Cato Institute" and "selfish politicians" as the enemy, with "community-owned renewable" energy as the solution to global warming.

People were more likely to agree with scientist's views about climate change after reading a story, rather than a list alone, regardless of which one they read.

"But what surprised us was how much the hero mattered," Jones said. People liked the villains less after reading the story, but that didn't affect their views much. Instead, having a hero they liked made them much more favorably disposed towards a solution. "Simple stories with likeable heroes are the most effective, they make people overlook incongruent things in the narrative," Jones says. "Obviously, this has implications across a lot of areas."

The findings don't mean that scientists suddenly need to invent parables to reach the public, he suggests, they just need to do more than just throw out the facts and hope that will do all the work. Instead of simply listing the evidence for climate change in reports, and then hoping people decide from hearing it that climate change is real, the findings suggest that scientists would be better off presenting their results in a narrative targeted to their audience's likes and dislikes. Libertarians, for example, might better listen to the facts about climate change if business is presented as the hero that can save the day from ill effects of increasing temperatures. Environmentalists want to hear about renewable energy. Normal folks (that's the hierarchical ones) will listen better if they hear that national security is threatened by a dependence on fossil fuels.

Of course, Jones acknowledges that some portion of people just won't accept the evidence for climate change no matter how it is presented, where about 12% of the population was "dismissive" of climate worries, according to a George Mason University survey released in June. "Some of the opposition to addressing climate change is completely rational," he adds, coming from regions of the country, such as West Virginia, where coal and oil interests would see prices in their industries rise with efforts to account for the environmental costs of the greenhouse gases created by burning fossil fuels.

The results aren't too surprising, says science writer Chris Mooney, who presented an American Academy of Arts & Sciences report, "Do Scientists Understand the Public?" this summer, looking at steps towards smarter public discussion of personal genomes, nuclear waste, energy and other new technologies. "Scientists have started taking steps in this direction," Mooney says, pointing to the National Academy of Sciences working with Hollywood writers. "They just need to take more."

Richard L. Itteilag

Scientists don't like to hear the story about telling stories, Jones adds. "One of the first places I presented this research was to scientists with the National Weather Service. They hated the idea that you have to tell people a story instead of just giving them 'the facts'," he says. "But the real question is do you want people to hear you, or not?"

Emerging Energy-Efficient Industrial Technologies

N. Martin, E. Worrell, M. Ruth, L. Price (LBNL)
R. N. Elliott, A. M. Shipley, J. Thorne (ACEEE)
October 2000

Executive Summary

U.S. industry consumes approximately 37 percent of the nation's energy to produce 24 percent of the nation's GDP. Increasingly, industry is confronted with the challenge of moving toward a cleaner, more sustainable path of production and consumption, while increasing global competitiveness. Technology will be essential for meeting these challenges. At some point, businesses are faced with investment in new capital stock. At this decision point, new and emerging technologies compete for capital investment alongside more established or mature technologies. Understanding the dynamics of the decision-making process is important to perceive what drives technology change and the overall effect on industrial energy use.

The assessment of emerging energy-efficient industrial technologies can be useful for:

- identifying R&D projects;
- identifying potential technologies for market transformation activities;
- providing common information on technologies to a broad audience of policy-makers; and
- offering new insights into technology development and energy efficiency potentials.

With the support of PG&E Co., NYSERDA, DOE, EPA, NEEA, and the Iowa Energy Center, staff from LBNL and ACEEE produced this assessment of emerging energy-efficient industrial technologies. The goal was to collect information on a broad array of potentially significant emerging energy-efficient industrial technologies and carefully characterize a sub-group of approximately 50 key technologies. Our use of the term "emerging" denotes technologies that are both pre-commercial but near commercialization, and technologies that have already entered the market but have less than 5 percent of current market share. We also have chosen technologies that are energy-efficient (i.e., use less energy than existing technologies and practices to produce the same product), and may have additional "non-energy benefits." These benefits are as important (if not more important in many cases) in influencing the decision on whether to adopt an emerging technology.

The technologies were characterized with respect to energy efficiency, economics, and environmental performance. The results demonstrate that the United States is not running out of technologies to improve energy efficiency and economic and environmental performance, and will not run out in the future. We show that many of the technologies have important non-energy benefits, ranging from reduced environmental impact to improved productivity and worker safety, and reduced capital costs.

Methodology

The assessment began with the identification of approximately 175 emerging energy-efficient industrial technologies through a review of the literature, international R&D programs, databases, and studies. The review was not limited to U.S. experiences, but rather we aimed to produce an inventory of international technology developments. We devised an initial screening process to select the most attractive technologies that had: (1) high potential energy savings; (2) lower comparative first costs relative to existing technologies; and (3) other significant benefits. While some technologies scored high on all of these characteristics, most had a mixed score. We formalized this approach in a very simple rating system. Based on the literature review and the application of initial screening criteria, we identified and developed profiles for 54 technologies. The technologies ranged from highly specific ones that can be applied in a single industry to more broadly crosscutting ones that can be used in many industrial sectors.

Each of the selected technologies has been assessed with respect to energy efficiency characteristics, likely energy savings by 2015, economics, and environmental performance, as well as what's needed to further the development or implementation of the technology. The technology characterization includes a one to two-page description and a one-page table summarizing the results for the technology.

Summary of Results

Table ES-1 provides an overview of the 54 emerging energy-efficient industrial technologies. We evaluated energy savings in two ways. The third column of Table ES-1 (Total Energy Savings) shows the amount of total manufacturing energy that the technology is likely to save in 2015 in a business-as-usual scenario. The fourth column (Sector Savings) reflects the savings relative to expected energy use in the particular sector. We believe that both metrics are useful in evaluating the relative savings potential of various technologies.

Economic evaluation of the technology is identified in the summary table by simple payback period, defined as the initial investment costs divided by the value of energy savings less any changes in operations and maintenance costs. We chose this measure since it is frequently used as a shorthand evaluation metric among industrial energy managers. Payback times for the technologies range from the immediate to 20 years or more. Of the 54 technologies profiled, 31 have estimated paybacks of 3 years or less, with six paying back immediately.

Energy savings are most often not the determining factor in the decision to develop or invest in an emerging technology. Over two-thirds of technologies not only save energy but yield non-energy benefits. We separated these non-energy benefits into environmental and other categories. We assessed how important the environmental benefits are to the technology adoption decision and listed the nature of the other benefit(s). We include an assessment of the importance of these non-energy benefits.

Technologies do not seamlessly enter existing markets immediately after development. The acceptance of emerging technologies is often a slow process that entails active research and development, prototype development, market demonstration, and other activities. In Table ES-1 we summarize the recommendations for the primary activities that could be undertaken to increase the technologies' rate of uptake. Over half of these technologies have already been developed to prototype stage or are already commercial but require further demonstration and dissemination.

Each technology is at a different point in the development or commercialization process. Some technologies still need further R&D to address cost or performance issues, some are ready for demonstration, and others have already proven themselves in the field and the market needs to be informed of the benefits and market channels needed to develop skills to deliver the technology. Our outlining of recommended actions in Table ES-1 is not an endorsement of any particular technology. Technology purchasers and users will ultimately decide regarding future development. However, the actions specified are intended to help identify whether a technology is both technically and economically viable and whether it is robust enough to accommodate the stringent product quality demands in various manufacturing establishments.

Seventeen emerging technologies could benefit from additional R&D. We suggest further R&D for several primary metal technologies, and several cross-cutting motor and utility technologies. In addition to private research funds, several of the identified technologies have received some R&D support from DOE or other public entities, including federal and state agencies.

There are also a large number of technologies that already have made some headway into the marketplace or are at the prototype testing stage, and therefore are candidates for demonstration for potential customers to gain comfort with the technology. While we recommend further demonstration and dissemination of these technologies, it was often difficult to understand what is limiting their uptake without more comprehensive investigation of market issues. Some of the technologies in this category are common in European countries or Japan but have not yet penetrated the U.S. market. Others are being newly developed in the United States and face challenges in reducing the risks perceived by potential purchasers. Two technologies, motor system optimization and pump efficiency improvement, are opportunities for training programs similar to those developed by DOE for the compressed air system management. For advanced industrial CHP turbine systems, the major recommended activity is removal of policy barriers. For other technologies, their unique markets will dictate the form of the educational and promotional activities. We urge the reader to follow up on any details in the specific technology profiles provided in Section VI of this report.

We assess the technology's likelihood of success in the marketplace. While our study evaluates each technology in relation to a given reference technology, the reality of the market is that technologies compete for market share. We made a judgement (based on the energy savings, cost-effectiveness, importance of non-energy benefits, market conditions, data reliability, and potential competing technologies) as to the likelihood that the technology would succeed in the marketplace.

From a national energy policy perspective, it is important to understand which technologies have both a high likelyhood of success and a high energy-savings. While various audiences may be interested in sector-specific or regional-specific technologies, the technologies listed in Table ES-2 are intended to provide guidance to those interested in the impact of energy-saving technologies on a more national level. This table also identifies the recommended next steps appropriate for each technology.

Conclusions and Recommendations for Future Work

For this study, we identified about 175 emerging energy-efficient technologies in industry, of which we characterized 54 in detail. While many profiles of individual emerging technologies are available, few reports have attempted to impose a standardized approach to the evaluation of the technologies. This study provides a way to review technologies in an independent manner, based on information on energy savings, economic, non-energy benefits, major market barriers, likelihood of success, and suggested next steps to accelerate deployment of each of the analyzed technologies.

There are many interesting lessons to be learned from further investigation of technologies identified in our preliminary screening analysis. The detailed assessments of the 54 technologies are useful to evaluate claims made by developers, as well as to evaluate market potentials for the United States or specific regions. In this report we show that many new technologies are ready to enter the market place, or are currently under development, demonstrating that the United States is not running out of technologies to improve energy efficiency and economic and environmental performance, and will not run out in the future. The study shows that many of the technologies have important non-energy benefits, ranging from reduced environmental impact to improved productivity. Several technologies have reduced capital costs compared to the current technology used by those industries. Non-energy benefits such as these are frequently a motivating factor in bringing technologies such as these to market.

Further evaluation of the profiled technologies is still needed. In particular, further quantifying the non-energy benefits based on the experience from technology users in the field is important. Interactive effects and intertechnology competition have not been accounted for and ideally should be included in any type of integrated technology scenario, for it may help to better evaluate market opportunities.

While this report focuses on the United States, state- or region-specific analysis of technologies may provide further insights into opportunities specific for the region served. Regional specificity is determined by the type of users (i.e., industrial activities) in the region, as well as the available technology developers. Combining region-specific circumstances with technology evaluations provided in this report may lead to recognition of varying needs and the appropriate policy choices for regional (e.g., state or utility) agencies.

Our selection of a limited set of 54 technologies was an arbitrary constraint based on the funding available. A number of the initial technologies screened appeared very interesting and warrant further study, but were eliminated due to resource constraints. In addition, the initial list of candidate technologies should not be viewed as all-encompassing. The authors are aware that other promising existing technologies exist, and that by their nature new technologies will be continually emerging. Ideally, the effort reflected in this report should be the start of a continuing process that identifies and profiles the most promising emerging energy-efficient industrial technologies and tracks the market success for these technologies. An interactive database may be a better choice for it would allow the continuous updating of information, rather than providing a static snapshot of the industrial technology universe.

6/30/09

NEW INDUSTRIAL ENERGY-EFFICIENCY TECHNOLOGIES AND THE EFFECTS ON INDUSTRIAL ENERGY INTENSITY (I.E., INDUSTRIAL CONSERVATION)

Part One

BY

RICHARD L. ITTEILAG
PRESIDENT
ENERGISTICS, INC.
A DIVISION OF WASHINGTON
PROPOSALASSOCIATES, INC.

Abstract

U.S. industry consumes over one-third of the nation's energy to produce a quarter of the nation's GDP. Increasingly, industry is confronted with the challenge of moving toward a cleaner, more sustainable path of production and consumption, while increasing global competitiveness. Innovative technologies are emerging and essential for meeting these challenges. At some point, businesses including users and technology manufacturers are faced with various investment decisions in new capital stock. At this decision point, new and emerging technologies, sometimes underutilized if at all adopted, often compete for capital investment alongside more established or mature technologies. The technologies in this report are characterized with respect to energy efficiency, economics, and environmental performance. The results demonstrate that the United States is not running out of technologies to improve energy efficiency and economic and environmental performance, and will not run out in the future. It is shown that many of the technologies have important non-energy benefits, ranging from reduced environmental impact to improved productivity and worker safety, and reduced capital costs.

Discussion

The United States (U.S.) industrial sector produces 24 percent of Gross Domestic Product (GDP) while consuming 37 percent of total U.S. energy supply. With over a third of the energy consumed at current output levels, U.S. industries are constrained to invest scarce capital on high-efficiency, emerging and commercially available, industrial technologies to control costs and remain competitive in an evolving global market. The American Council on an Energy-Efficient Economy (ACE3) found 54 key industrial energy technologies both emerging and commercially available that will significantly reduce energy consumption per unit of output *permanently* across all major industrial processes and all major industry sectors. These technologies range, from industry-specific types applying to only one industry, to a broad cross-cutting array of technologies that can be used in many industrial sectors. The industrial energy-intensity landscape is now forever transformed to levels *dramatically* lower than previously experienced.

To achieve energy and peak demand reductions at industrial facilities, savings are achieved through the systematic evaluation of electric energy-using systems and subsequent implementation of energy-efficiency measures.

Energy Reduction is a Systematic Two-Step Process

Step 1. Evaluation

How does your facility rate? Inefficient energy asset equipment, nonoperational control strategies, faulty equipment, and deferred maintenance may result in overall system inefficiencies that are not readily noticeable. As a result, facility owners may not be aware of the sizeable savings opportunities that are available to them. The Targeted Industrial Energy Efficiency Program will provide an engineering assessment regarding the condition of the energy assets at your facility and identify opportunities to generate energy savings and improve overall operating performance. The result of the audit will be a written report that details recommended ways to save energy and improve the efficiency of your facility's systems.

Step 2. Implementation

Ready to improve your facility's energy efficiency? After reviewing the audit report, select your own contractor or utilize in-house staff to implement the recommended energy efficiency measures. The Targeted Industrial Energy Efficiency Program will provide financial incentives to buy down the cost of installation and implementation of approved energy efficiency measures.

For a typical industrial facility, approximately 10% of the electricity consumed is for generating compressed air. Optimization of compressed air systems can provide energy-efficiency improvements of 20 to 50 percent.

Emerging and Commercially Available Industrial Technologies

1. Refrigeration/Cooling (food/beverages): Energy consumed to freeze, cool and store meat, fish, fruit, vegetables, beverages as well as frozen products (e.g., ice cream, juices, etc.) is primarily performed by compressors powered by electricity. Some 70 trillion Btus were consumed in 2002

in the United States for cooling and refrigeration (i.e., 90% electricity) according to the U.S. Department of Energy.
2. Natural Gas Engines: Instead of an electric motor, a natural gas engine could drive a compressor. The gas engine is used as a direct drive and is considerably more energy efficient because the gas engine can follow the refrigeration loads quite accurately by utilizing a variable-speed drive engine not possible with an electrically drive compressor. In addition, waste heat off the engine can be used for space or water heating. This system generates large energy savings and reduces peak electricity consumption, as well as the resultant on-peak kW charges/on-peak electricity kWh charges. Energy savings can reach 52 to 77 percent.[1]

Conclusion

Energy Policy Act of 2005 (EPACT of 2005)

The comprehensive array of industrial conservation alternatives that are found in the

Energy Policy Act of 2005 (EPACT of 2005) including smart meters are as follows:

a. DDC (direct-drive controls)
b. "Green" buildings
c. Demand response
d. Real-time pricing
e. Temperature setbacks
f. Ventilation control
g. Boiler optimization

---

[1] A future White Paper, **Part Two**, will contain the 'many' additional emerging and commercially available industrial technologies by industry and industrial end-use including the expected energy savings.

h. Lighting products/systems
i. Smart meters
g. Reflective roof coatings
k. Cold-water detergents
l. Radiant-heated flooring
m. Concrete construction material
n. LED (light-emitting diodes) lights

7/31/09

NEW INDUSTRIAL ENERGY-EFFICIENCY TECHNOLOGIES AND THE EFFECTS ON INDUSTRIAL ENERGY INTENSITY (I.E., INDUSTRIAL CONSERVATION)

Part Two

BY

RICHARD L. ITTEILAG
PRESIDENT
ENERGISTICS, INC.
A DIVISION OF WASHINGTON
PROPOSALASSOCIATES, INC.

Abstract

The United States (U.S.) industrial sector has available for its purchase in the 21st century some 50-plus high-efficiency technologies that will forever change the industrial energy-intensity landscape, and environmental footprint, as well for generations at a markedly lower level of energy utilization and absolute consumption of Btus, on- and off-peak.

The technologies in this report are characterized with respect to energy efficiency, economics, and environmental performance. The results demonstrate that the United States is not running out of technologies to improve energy efficiency, economic growth and environmental performance, and will not run out in the future. It is shown that many of the technologies have important non-energy benefits, ranging from reduced environmental impact to improved productivity, worker safety and reduced capital costs.

Discussion

The American Council on an Energy-Efficient Economy (ACE3) found 54 key industrial energy technologies both emerging and commercially available that will significantly reduce energy consumption per unit of output *permanently* across all major industrial processes and all major industry sectors. These technologies range, from industry-specific types applying to only one industry, to a broad cross-cutting array of technologies that can be used in many industrial sectors. The industrial energy-intensity landscape is now forever transformed to levels *dramatically* lower than previously experienced.

To achieve energy and peak demand reductions at industrial facilities, here are 54 natural gas and electric energy-using systems, as well as subsequent implementation of energy-efficiency measures:

Emerging and Commercially Available Industrial Technologies

3. Refrigeration/Cooling (food/beverages): Energy consumed to freeze, cool and store meat, fish, fruit, vegetables, beverages as well as frozen products (e.g., ice cream, juices, etc.) is primarily performed by compressors powered by electricity. Some 70 trillion Btus were consumed in 2002 in the United States for cooling and refrigeration (i.e., 90% electricity) according to the U.S. Department of Energy.
4. Natural Gas Engines: Instead of an electric motor, a natural gas engine could drive a compressor. The gas engine is used as a direct drive and is considerably more energy efficient because the gas engine can follow the refrigeration loads quite accurately by utilizing a variable-speed drive engine not possible with an electrically drive compressor. In

addition, waste heat off the engine can be used for space or water heating. This system generates large energy savings and reduces peak electricity consumption, as well as the resultant on-peak kW charges/on-peak electricity kWh charges. Energy savings can reach 52 to 77 percent.
5. Thermal Storage: The use of off-peak electricity in the food industry to produce ice which is stored in ice pods or ice tanks. Energy savings can reach as high as 80 percent.
6. Emerging Refrigerants: Alternative working fluids to chlorofluorocarbons (CFCs) hydrofluorocarbons (HCFCs). These alternative working fluids such as lithium bromide (libr) can save 2 percent to 20 percent of the traditional electrical energy consumed.
7. Aluminum:

Conclusion

The United States (U.S.) industrial sector has available for its purchase in the 21st century some 50-plus high-efficiency technologies that will forever change the industrial energy-intensity landscape, and environmental footprint, as well for generations at a markedly lower level of energy utilization and absolute consumption of Btus, on-and off-peak.

It is shown that many of these technologies have important non-energy benefits, ranging from reduced environmental impact to improved productivity, worker safety and reduced capital costs. Conservatively, the implementation of must or all of these technologies over the next 20 years will generate an energy savings in the 15 percent-20 percent range, i.e., an overall conservation rate in the industrial sector of nearly 20 percent.

Industrial Technologies (Oct. 2010)

Technologies

Industry is the largest and most diverse energy-consuming sector in the United States, but it is often unable to accept the risks associated with the capital-intensive technology development required to decrease its energy use. The Industrial Technologies Program (ITP) can help to mitigate that risk. By supporting public-private partnerships, we bring together the strengths of business and government to meet the challenges of improving energy efficiency.

ITP and our industry partners have created an effective model for technology development. First, industry defines a vision and long-term goals. Next, we work together to build technology roadmaps with specific research and development pathways. Finally, ITP and industry share the costs of research and development (R&D) projects specified by those roadmaps and reach the common goals of improving not only energy efficiency, but economic viability, energy security, environmental quality, and resource conservation. ITP's efforts have resulted in over 160 technologies successfully reaching the marketplace, providing significant economic and environmental impacts for the United States.

Energy Intensive Industries

The specific industrial process required often determines energy use in industry. For example, the aluminum industry uses large amounts of electricity for smelting while the glass industry uses large amount of natural gas to melt silica in furnaces. With our Energy Intensive Industries process, ITP focuses specific R&D on the nation's eight most energy-intensive industries—aluminum, chemicals, forest product, glass, metal casting, mining, petroleum refining, and steel. Through the development of new technologies and processes in the Energy Intensive Industries, we expect to save 2.0 quads of energy and $7.9 billion, and to avoid 36.7 million metric tons carbon equivalent (MMTCE) of climate change gases by 2020.

- Aluminum
- Chemicals
- Forest Products
- Glass
- Metal Casting
- Mining
- Petroleum Refining
- Steel

Crosscutting Technologies

Crosscutting technologies are common to most manufacturing processes, and have widespread benefits to many different industries. Because these crosscutting technologies are so widely used, a small improvement in efficiency can yield large energy savings across many industries. By performing R&D on crosscutting technologies, and implementing successful results, ITP estimates industry could save .8 quads of energy and $3.4 billion, and avoid 13.2 MMTCE of climate change gases by 2020.

- Combustion
- Distributed Energy
- Energy Intensive Processes
- Fuel & Feedstock Flexibility
- Industrial Materials for the Future
- Nanomanufacturing
- Sensors & Automation

Emerging Technologies

As a measure of the success of the Energy Intensive Industries and crosscutting R&D strategies, ITP currently has more than 120 technologies that are emerging from research and development and are expected to be ready for commercialization within the next one to two years. Of these, more than 50 have been identified as being immediately ready for field testing.

ITP Commercial Successes

ITP tracks energy savings as well as other benefits associated with the successfully commercialized technologies resulting from its research partnerships. Our *Impacts* report summarizes some of these benefits including energy savings, waste reduction, increased productivity, lowered carbon dioxide and air pollutant emissions, and improved product quality.

Other EERE Programs

The Industrial Technologies Program is integral to the Office of Energy Efficiency and Renewable Energy (EERE) and works hand in hand with other EERE sectors. Our industrial technologies can benefit from, and contribute to, research and development efforts across EERE. For a complete list of other EERE Programs, visit the EERE Home Page.

6/30/10

NUCLEAR POWER GENERATION: CAPITAL, OPERATING AND ENDUSER COSTS

BY

RICHARD L. ITTEILAG
PRESIDENT
ENERGISTICS, INC.
A DIVISION OF WASHINGTON PROPOSAL
ASSOCIATES, INC.

Abstract

Currently, nuclear power is touted as a highly attractive source of new electric generation capacity in the United States (U.S.). However, in a competitive energy market, and particularly in tough economic times, the least cost strategy for building essential new generation capacity must be employed. This paper discusses the fundamental components of nuclear power plant costs.

Discussion

The U.S. residential, commercial and industrial sectors are bracing for new electric generation capacity and the resultant increased costs in electricity prices associated with this new capacity. One of the prime new capacity options is nuclear power generation. While no new nuclear plant has been constructed in the last 30+years in

the U.S., a number of electric utilities are planning new plants. In order to receive approval for these
facilities by the appropriate Public Utility Commissions, the electric companies must prepare detailed cost estimates for these planned plants. The components of nuclear energy cost production do not contribute equally to the final cost to the end user. Construction costs represent the largest portion ranging from about 65-75% of total cost, followed by operating costs at about 23-33% and finally decommissioning costs which contribute a meager 2-3% to the total.

Construction Costs

The majority of nuclear energy's cost to the consumer is the extensive capital investment required to construct the plant. Construction costs for a nuclear plant can be divided into several subcomponents.

Real Estate: Cost of the land for the plant and the requisite area to surround the plant for security.

Materials: Commodities such as copper, steel, and concrete that go in to the construction of the plant.

Engineering and Design: The costs of engineers and contractors to design and assemble the plant.

Licensing: Costs incurred with inspecting and acquiring approval from the Nuclear Regulatory Commission (NRC) for operating the plant.

Financing: Cost of borrowing capital to build the plant.

Projections for new nuclear construction range dramatically due to fluctuations in materials, financing costs, the uncertainty of engineering and design costs, and the inability to accurately forecast the benefit of economies of volume. The nuclear power plants that exist in the United States today were all built with unique and varying designs. Current proposals for new nuclear construction in the U.S.

revolve around the idea of implementing a standardized reactor design that would result in manufacturing synergies from utilization of standard parts, quicker approval from the NRC, and reduction in overall completion time due to the practice effect.

Operating Costs

Employee: Payroll, benefits, and other employee-related expenses make up the largest portion of operating costs.

Supplies: Commodities used in activities required to run the plant including items such as water, industrial gases, office supplies, gloves, paper, uniforms, etc.

Administration: Nuclear power plants are owned by companies that require administrative functions such as Human Resources, Information Technology, Legal, and Accounting to fulfill their obligations as a business.

Fuel: The fuel costs to produce energy from a nuclear reactor.

The Operating costs of a nuclear power plant are the daily expenses incurred to operate both the plant itself and the business that owns it. Operating costs are the second largest contributor of cost to the consumer.

Waste Disposal

Nuclear energy produces no emissions so the waste from energy production does not simply disperse into the atmosphere. The nuclear reaction produces waste that is then disposed of in dry casks and stored at facilities located across the United States and the world. Nuclear energy production results in a small volume of waste that is inexpensive to dispose. There is the possibility of recycling nuclear spent fuel, but that method would only be cost effective if economies of volume in third generation nuclear reactor construction were achieved. Nuclear spent fuel recycling does have the benefit of reducing the operating cost of fuel, but that would be offset by the cost of the recycling facility. When analyzing the cost structure of nuclear energy, the cost of waste disposal is actually included in the fuel subcomponent of operating costs.

Decommissioning

Companies that operate nuclear power plants are required to set aside money each year to build a reserve of funds to pay for the decommissioning of the plant at the end of its useful life. The projected cost to decommission the plant is calculated and spread across the expected life of the plant. Nuclear plants are projected to have a useful life of 40 years with the ability to extend that lift through an application and review process with the NRC.

Conclusion

The components of nuclear energy cost production do not contribute equally to the final cost to the end user. Construction costs represent the largest portion ranging from about 65-75% of total cost, followed by operating costs at about 23-33% and finally decommissioning costs which contribute a meager 2-3% to the total.

September, 2009 White Paper

9/30/09

"POTENTIAL CONSERVATION CREATED BY NATIONAL DEMAND RESPONSE PROGRAMS ON ELECTRICITY DEMAND CAPACITY"[1]

Introduction

The increased use of demand response capacity has driven electricity demand, and in turn prices, down. As customers utilized more electricity under demand response programs, they actually reduce their average consumption markedly. The incentive for customers to conserve is the significant revenue they earn under the requirements of demand response program pricing. Weak electricity prices are highly correlated to significant increases in electricity demand response capacity where C/I customers receive payments for reductions in consumption, i.e., conserving electricity use. This current trend begs the fundamental question: how the growth in demand response programs will affect peak electricity demand over the next 10 years on a national basis?

---

[1] A future White Paper will discuss the regional results of these four cases by the nine Census Regions.

Discussion

According to a Federal Energy Regulatory Commission (FERC) report dated June, 2009 entitled, A National Assessment of Demand Response Potential, U.S. peak electricity capacity will grow 1.7% per year through 2019 reaching above 950 GW from a base of 810 GW in 2009 without any demand response programs instituted over this 10-year timeframe. These estimates were provided to FERC by the North American Electric Reliability Council (NERC). The C/I customers in this capacity sector are highly price sensitive. Pricing is based on usage patterns and adjusts on a real-time basis determined by "smart meters." The more these C/I customers conserve electricity, the more revenue they earn and, therefore, the more effective these programs are in reducing electricity consumption, i.e, promoting conservation.

In order to assess the future impact, if any, on national peak electricity capacity, FERC postulated four scenarios: 1) Business-as-Usual (BAU), the starting point or benchmark case where current and planned demand response programs are constant; 2) Extended BAU (EBAU), the case where the current mix of demand response programs are expanded to all states and achieves 'best practices' levels of participation, including pricing program incentives; 3) Achievable Participation (AP), where smart meters are deployed and dynamic pricing or incentive/discount pricing is the default tariff; and 4) Full Participation (FP), the potential case where the total amount of cost-effective demand response programs are employed.

Conclusions

The results in the form of conservation or reduction in peak demand electricity capacity from the NERC case are as follows:

1. Peak electricity capacity in the BAU case grows at a rate of 1.7% per year, similar to the NERC case, but reaches a lower level of peak electricity demand capacity. The reduction in peak electricity demand capacity is 37 GW to 773 GW of

total capacity in 2009 and the reduction or conservation by the year 2019 is 38 GW or 912 GW of total peak electricity capacity. This case produces a conservation rate of 4%.
2. The EBAU case produces a peak electricity demand capacity growth rate of 1.3 % per year and a reduction in peak electricity demand capacity of 82 GW by 2019, or a 9% level of conservation. Total electricity demand capacity reaches 868 GW in 2019.
3. The AP case achieves an even greater reduction in peak electricity demand capacity by 2019, as electricity demand on peak grows at only 0.6 %. Total electricity demand capacity on peak by 2019 is 812 GW, or a reduction of 14% for conservation due to the widespread implementation of demand response programs across the entire U.S.
4. The FP case produces the largest reduction in peak electricity demand capacity or level of conservation. In this case, peak electricity demand capacity growth is flat or 0%, and total peak capacity reaches a modest 762 GW creating a conservation rate of 20%. This is a reduction in peak electricity capacity of 188 GW.

In general, demand response programs, when implemented either on a sporadic bases in the some states or, more significantly, introduced and utilized on a widespread basis to their fullest in every state, will achieve a considerable reduction in peak electricity demand capacity, or significant levels of conservation, without 'command-and-control' federal/state government mandates.

9/30/09

"POTENTIAL CONSERVATION CREATED BY REGIONAL DEMAND RESPONSE ELECTRICITY CAPACITY"

The increased use of demand response capacity has driven electricity demand, and in turn prices, down. As customers utilized more electricity under demand-response programs, they actually reduce their average consumption markedly. The incentive for customers to conserve is the significant revenue they earn under the requirements of demand response program pricing.

Weak electricity prices are highly correlated to significant increases in electricity demand response capacity where C/I customers receive payments for reductions in consumption, i.e., conserving electricity use. The Federal Energy Regulatory Commission (FERC) report dated August, 2009, estimates that there is some x amount of electricity capacity classified as demand-response, y MW, of z percent of total U.S. electricity capacity. The C/I customers in this capacity sector are highly price sensitive. Pricing is based on usage patterns and adjusts on a real-time basis determined by "smart meters." The more these C/I customers conserve electricity, the more revenue they earn demand-response programs and believed that these fairly new and novel programs were effective in reducing electricity consumption, i. e, promoting conservation.

In this FERC report, a National Assessment of Demand Response Potential which is required by the Energy Independence and Security Act of 2007 (EISA 2007), available at the Conference, some 150 GW of peak electricity demand in the U.S. could be reduced by demand response programs by 2019 compared to a business as usual future without these programs.

The main incentive for commercial and industrial customers to participate in these demand-response programs is the revenue gained from demand response pricing.

The Conference was sponsored by the Demand Response Coordinating Committee. The definition of demand response is providing electricity customers, whether retail or wholesale, a price choice determined by time-based pricing through smart meters to reduce consumption or shift consumption off-peak in return for price compensation, Most recent program activities at several utilities have focused on the use of pricing and other incentives to involve customers in demand response programs. The operative term in demand response programs is: *incentives*.

Another area discussed in depth and correlated closely with the demand response capacity is the "smart grid."

While planning and R&D has accelerated, implementation is only a near-term probability. The many different technologies that come into play in the delivery of demand response and smart grid programs/activities include such areas as: information display, distributed storage and distributed generation (DR). One of the most widespread areas of demand response activities for commercial and industrial customers is DG, particularly in the context of the smart grid. Another end-use area that is becoming a dynamic demand response and smart grid application is street lighting. In addition, plug-in hybrid vehicles will be integrated into the smart grid.

The DRCC Mission is:

- To increase the knowledge base in the U.S. on demand response
- To facilitate the exchange of information and expertise about demand response among interested parties
- To build and establish a demand response "community" of policy makers, utilities, system operators, technology companies and other stakeholders.

The DRCC Members are:

American Electric Power (AEP)
Arizona Public Service
DTE Energy
ISO New England
Landis+Gyr
MidAmerican Energy (PG&E)
Midwest ISO
National Grid
NYSERDA
Pacific Gas & Electric (PG&E)
PJM Interconnection
Progress Energy
Salt River Project
San Diego Gas & Electric (SDG&E)
Southern California Edison
Tenesse3e Valley Authority
Viridity Energy
Wal-Mart
Xcel Energy

For more information on the previous Town Hall Meetings held by DRCC, please refer to the DRCC Web site: demandresponsecommittee.org.

Richard L. Itteilag

Smart Grid Standards (Oct. 2010)

A group of smart grid industry players which include big names such as Pacific Gas & Electric Company, Southern California Edison, Honeywell, and the Lawrence Berkeley National Laboratory (Berkeley Lab) recently announced the formation of the OpenADR Alliance, a non-profit organization created to advocate the development, adoption and compliance of a smart grid standard called Open Automated Demand Response (OpenADR). The alliance's main goal is to reduce costs, improve reliability and accelerate the speed of Automated Demand Response (Auto-DR) and smart grid implementations in the US.

Through an automated message delivery sent by utilities to consumers, Auto-DR suggests households and business establishments to reduce usage of electricity during peak demand hours, or in reaction to market price changes. What OpenADR does is that it standardizes a message format for the Auto-DR to ensure that dynamic price and reliability signals are sent in a uniform and interoperable data model among utilities, Independent System Operators (ISO) and customers' energy management and control systems.

In specific, utilities, vendors, and consumers will benefit from the adoption of OpenADR as it brings cost reduction (mostly production, service, and maintenance costs due to the standardization), compliance assurance (through a uniform set of standards to which vendor technologies need to adhere to), and improved reliability (as a result of standardizing a message format for Auto-DR).

The OpenADR Alliance's role will be to initiate collaboration between industry stakeholders in order to ensure the rapid development of OpenADR. National standards work will be based from OpenADR specifications published by Lawrence Berkeley National Laboratory and will be funded by the California Energy Commission's Public Interest Energy Research (PIER) program. Also, the National Institute of Standards and Technology's (NIST) is further developing

OpenADR, along with other organizations such as the Organization for the Advancement of Structured Information Standards (OASIS), the Utilities Communications Architecture International User's Group (UCAIug), and the North American Energy Standards Board (NAESB).

According to Mary Ann Piette, research director for PIER Demand Response Research Center (DRRC) at Berkeley Lab, "Grounded in the standards activities initiated by Berkeley Lab in 2002, the OpenADR Alliance will play a central role in accelerating the adoption of Auto-DR and rapidly advancing our power grid into the 21st century". "Only through interoperable technology standards can we implement Smart Grid solutions with the reliability, cost-effectiveness and guaranteed compliance necessary for broad market acceptance. The OpenADR Alliance will implement processes to quickly bring this commercially proven standard to market", she added.

Meanwhile, Honeywell's vice president of energy solutions, Jeremy Eaton said,"There's no question the widespread adoption of an OpenADR standard will lower the development, equipment and service costs for Smart Grid vendors and the utilities investing in these solutions". "And it will ultimately benefit homeowners and businesses because open standards spur competition and innovation, and will lead to more effective Smart Grid technologies, and greater energy and cost savings."

Solar Generation Description of Project (Oct. 2010)

General Description of Project

On August 31, 2007, Solar Partners I, LLC, Solar Partners II, LLC, Solar Partners IV, LLC and Solar Partners VIII, LLC (Solar Partners) submitted a single Application for Certification (AFC) to the California Energy Commission to develop three solar thermal power plants and shared facilities in close proximity to the Ivanpah Dry Lake, in San Bernardino County, California on federal land managed by the Bureau of Land Management (BLM). The proposed project would be constructed in three phases: two 100-megawatt (MW) phases (known as Ivanpah 1 and Ivanpah 2) and a 200-MW phase (Ivanpah 3). The three plants are collectively referred to as the Ivanpah Solar Electric Generating System (ISEGS) and would be located in:

- Southern California's Mojave Desert, near the Nevada border, to the west of Ivanpah Dry Lake
- San Bernardino County 4.5 miles southwest of Primm, Nevada, 3.1 miles west of the California-Nevada border
- Township 17N, Range 14E, and Township 16N, Range 14E

Project Description

The proposed project includes three solar concentrating thermal power plants, based on distributed power tower and heliostat mirror technology, in which heliostat (mirror) fields focus solar energy on power tower receivers near the center of each heliostat array. Each 100-MW site would require approximately 850-acres (or 1.3 square miles) and would have three tower receivers and arrays; the 200-MW site would require approximately 1,600-acres (or 2.5 square miles) and would have 4 tower receivers and arrays. The total area required for all three phases would including the administration building/operations and maintenance building and substation and be approximately 3,400-acres (or 5.3 square miles). Given that the three plants would be developed in concert, the proposed solar plant projects would share the common facilities mentioned above to include access roads, and the reconductored transmission lines for all three phases. Construction of the entire project is anticipated to begin in the first quarter of 2009, with construction being completed in the last quarter of 2012.

Process Description

In each solar plant, one Rankine-cycle reheat steam turbine receives live steam from the solar boilers and reheat steam from one solar reheater located in the power block at the top of its own tower. The reheat tower would be located adjacent to the turbine. Additional heliostats would be located outside the power block perimeter road, focusing on the reheat tower. Final design layout locations are still being developed. The solar field and power generation equipment would be started each morning after sunrise and insolation build-up, and shut down in the evening when insolation drops below the level required to keep the turbine online.

Each plant also includes a partial-load natural gas-fired steam boiler, which would be used for thermal input to the turbine during the morning start-up cycle to assist the plant in coming up to operating temperature more quickly. The boiler would also be operated during transient cloudy conditions, in order to maintain the turbine on-line and ready to resume production from solar thermal input, after the clouds pass. After the clouds pass and solar thermal input resumes, the turbine would be returned to full solar production. Each plant uses an air-cooled condenser or "dry cooling," to minimize water usage in the site's desert environment. Water consumption would therefore, be mainly to provide water for washing heliostats. Auxiliary equipment at each plant includes feed water heaters, a deaerator, an emergency diesel generator, and a diesel fire pump.

Electricity would be produced by each plant's Solar Receiver Boiler and the steam turbine generator. The heliostat mirrors would be arranged around each solar receiver boiler. Each mirror tracks the sun throughout the day and reflects the solar energy to the receiver boiler. The heliostats would be 7.2-feet high by 10.5-feet wide (2.20-meters by 3.20-meters) yielding a reflecting surface of 75.6

square feet (7.04 square meters). They would be arranged in arcs around the solar boiler towers asymmetrically.

Each solar development phase would include:

- a natural gas-fired start-up boiler to provide heat for plant start-up and during temporary cloud cover;
- an air-cooled condenser or "dry cooling," to minimize water usage in the site's desert environment;
- one Rankine-cycle reheat steam turbine that receives live steam from the solar boilers and reheat steam from one solar reheater located in the power block at the top of its own tower adjacent to the turbine; and
- a raw water tank with a 250,000 gallon capacity; 100,000 gallons to be used for the plant and the remainder to be reserved for fire water.
- a small onsite wastewater plant located in the power block that treats wastewater from domestic waste streams such as showers and toilets;
- auxiliary equipment including feed water heaters, a deaerator, an emergency diesel generator, and a diesel fire pump.

Transmission

Ivanpah 1, 2 and 3 would be interconnected to the Southern California Edison (SCE) grid through upgrades to SCE's 115-kV line passing through the site on a northeast-southwest right-of-way. Upgrades would include a new 220/115-kV breaker and-a-half substation between the Ivanpah 1 and 2 project sites. The existing 115-kV transmission line from the El Dorado substation would be replaced with a double-circuit 220-kV overhead line that would be interconnected to the new substation. Power from Ivanpah 1, 2 and 3 would be transmitted at 115-kV to the new substation.

Natural Gas

Natural gas supply for ISEGS would connect to the Kern River Gas Transmission Company (KRGT) pipeline about 0.5 miles north of the Ivanpah 3 site.

Water Use and Discharge

Raw ground water would be drawn from one of two wells, located east of Ivanpah 2, which would provide water to all three plants. Each well would have sufficient capacity to supply water for all three phases. Actual water is not expected to exceed 100 afy for all three plants. Groundwater would go through a treatment system for use as boiler make-up water and to wash the heliostats. No wastewater would be generated by the system, except for a small stream that would be treated and used for landscape irrigation.

Energy Commission and Bureau of Land Management Joint Review Process

The BLM and the Energy Commission have executed a Memorandum of Understanding concerning their intent to conduct a joint environmental review of all three plants in a single National Environmental Policy Act (NEPA)/California Environmental Quality Act (CEQA) process. It is in the interest of the BLM and the Energy Commission to share in the preparation of a joint environmental analysis of the proposed project to avoid duplication of staff efforts, to share staff expertise and information, to promote intergovernmental coordination at the local, state, and federal levels, and to facilitate public review by providing a joint document and a more efficient environmental review process.

Under federal law, the BLM is responsible for processing requests for rights-of-way to authorize the proposed project and associated transmission lines and other facilities to be constructed and operated on land it manages. In processing applications, the BLM must comply with the requirements of NEPA, which requires that federal agencies reviewing projects under their jurisdiction consider the environmental impacts associated with the proposed project construction and operation.

As the lead agency under CEQA, the Energy Commission is responsible for reviewing and ultimately approving or denying all applications to construct and operate thermal electric power plants, 50 MW and greater, in California. The Energy Commission's facility certification process carefully examines public health and safety, environmental impacts and engineering aspects of proposed power plants and all related facilities such as electric transmission lines and natural gas and water pipelines.

The first step in the Energy Commission's review process is for staff to determine whether or not the AFC contains all the information required by its regulations. When the Energy Commission determines the AFC is complete, staff will begin data discovery and issue analysis phases. At that time, a detailed examination of the issues will occur.

Energy Commission Facility Certification Process

The Energy Commission's facility certification process carefully examines public health and safety, environmental impacts and engineering aspects of proposed power plants and all related facilities such as electric transmission lines, natural gas pipelines, etc. The Energy Commission is the lead agency under the California Environmental Quality Act (CEQA) and has a certified regulatory program under CEQA. Under its certified program, the Energy Commission is exempt from having to prepare an environmental impact report. Its certified program, however, does require environmental analysis of the project, including an analysis of alternatives and mitigation measures to minimize any significant adverse effect the project may have on the environment.

Solar power

Solar power is the conversion of sunlight into electricity, either directly using photovoltaics (PV), or indirectly using concentrated solar power (CSP). Commercial CSP plants were first developed in the 1980s, and the 354 MW SEGS CSP installation is the largest solar power plant in the world and is located in the Mojave Desert of California. The 80 MW Sarnia Photovoltaic Power Plant in Canada, is the world's largest photovoltaic plant. Experimental approaches to solar power include concentrated photovoltaics systems, thermovoltaic devices, and space-based solar power.

Solar power is an intermittent energy source, meaning that solar power is not available at all times, and is normally supplemented by storage or another energy source, for example with wind power and hydropower.

Applications

Average insolation showing land area (small black dots) required to replace the world primary energy supply with solar electricity. 18 TW is 568 Exajoule (EJ) per year. Insolation for most people is from 150 to 300 W/m² or 3.5 to 7.0 kWh/m²/day.

Solar power is the conversion of sunlight into electricity. Sunlight can be converted directly into electricity using photovoltaics (PV), or indirectly with concentrated solar power (CSP), which normally focuses the sun's energy to boil water which is then used to provide power, and other technologies, such as the sterling engine dishes which use a sterling cycle engine to power a generator. Photovoltaics were initially used to power small and medium-sized applications, from the calculator powered by a single solar cell to off-grid homes powered by a photovoltaic array.

The only significant problem with solar power is installation cost, although cost has been decreasing due to the learning curve. Developing countries in particular may not have the funds to build solar power plants, although small solar applications are now replacing other sources in the developing world.

Concentrating solar power

Concentrating Solar Power (CSP) systems use lenses or mirrors and tracking systems to focus a large area of sunlight into a small beam. The concentrated heat is then used as a heat source for a conventional power plant. A wide range of concentrating technologies exists; the most developed are the parabolic trough, the concentrating linear fresnel reflector, the Stirling dish and the solar power tower. Various techniques are used to track the Sun and focus light. In all of these systems a working fluid is heated by the concentrated sunlight, and is then used for power generation or energy storage.

A parabolic trough consists of a linear parabolic reflector that concentrates light onto a receiver positioned along the reflector's focal line. The receiver is a tube positioned right above the middle of the parabolic mirror and is filled with a working fluid. The reflector is made to follow the Sun during the daylight hours by tracking along a single axis. Parabolic trough systems provide the best land-use factor of any solar technology. The SEGS plants in California and Acciona's Nevada Solar One near Boulder City, Nevada are representatives of this technology. The Suntrof-Mulk parabolic trough, developed by Melvin Prueitt, uses a technique inspired by Archimedes' principle to rotate the mirrors.

Concentrating linear fresnel reflectors are CSP-plants which use many thin mirror strips instead of parabolic mirrors to concentrate sunlight onto two tubes with working fluid. This has the advantage that flat mirrors can be used which are much cheaper than parabolic mirrors, and that more reflectors can be placed in the same amount of space, allowing more of the available sunlight to be used. Concentrating linear fresnel reflectors can be used in either large or more compact plants.

A stirling solar dish, or dish engine system, consists of a stand-alone parabolic reflector that concentrates light onto a receiver positioned at the reflector's focal point. The reflector tracks the Sun along two axes. Parabolic dish systems give the highest efficiency among CSP technologies. The 50 kW Big Dish in Canberra, Australia is an example of this technology. The stirling solar dish combines a parabolic concentrating dish with a stirling heat engine which normally drives an electric generator. The advantages of stirling solar over photovoltaic cells are higher efficiency of converting sunlight into electricity and longer lifetime.

A solar power tower uses an array of tracking reflectors (heliostats) to concentrate light on a central receiver atop a tower. Power towers are more cost effective, offer higher efficiency and better energy storage capability among CSP technologies. The Solar Two in Barstow, California and the Planta Solar 10 in Sanlucar la Mayor, Spain are representatives of this technology.

Photovoltaics

A solar cell, or photovoltaic cell (PV), is a device that converts light into electric current using the photoelectric effect. The first solar cell was constructed by Charles Fritts in the 1880s. Although the prototype selenium cells converted less than 1% of incident light into electricity, both Ernst Werner von Siemens and James Clerk Maxwell recognized the importance of this discovery. Following the work of Russell Ohl in the 1940s, researchers Gerald Pearson, Calvin Fuller and Daryl Chapin created the silicon solar cell in 1954. These early solar cells cost 286 USD/watt and reached efficiencies of 4.5-6%.

Solar power has great potential, but in 2008 supplied less than 0.02% of the world's total energy supply[citation needed]. There are many competing technologies, including fourteen types of photovoltaic cells, such as thin film, monocrystalline silicon, polycrystalline silicon, and amorphous cells, as well as multiple types of concentrating solar power. It is too early to know which technology will become dominant.

The earliest significant application of solar cells was as a back-up power source to the Vanguard I satellite in 1958, which allowed it to continue transmitting for over a year after its chemical battery was exhausted. The successful operation of solar cells on this mission was duplicated in many other Soviet and American satellites, and by the late 1960s, PV had become the established source of power for them. Photovoltaics went on to play an essential part in the success of early commercial satellites such as Telstar, and they remain vital to the telecommunications infrastructure today.

The high cost of solar cells limited terrestrial uses throughout the 1960s. This changed in the early 1970s when prices reached levels that made PV generation competitive in remote areas without grid access. Early terrestrial uses included powering telecommunication stations, off-shore oil rigs, navigational buoys and railroad crossings. These off-grid applications accounted for over half of worldwide installed capacity until 2004.

The 1973 oil crisis stimulated a rapid rise in the production of PV during the 1970s and early 1980s. Economies of scale which resulted from increasing production along with improvements in system performance brought the price of PV down from 100 USD/watt in 1971 to 7 USD/watt in 1985. Steadily falling oil prices during the early 1980s led to a reduction in funding for photovoltaic R&D and a discontinuation of the tax credits associated with the Energy Tax Act of 1978. These factors moderated growth to approximately 15% per year from 1984 through 1996.

Since the mid-1990s, leadership in the PV sector has shifted from the US to Japan and Europe. Between 1992 and 1994 Japan increased R&D funding, established net metering guidelines, and introduced a subsidy program to encourage the installation of residential PV systems. As a result, PV installations in the country climbed from 31.2 MW in 1994 to 318 MW in 1999, and worldwide production growth increased to 30% in the late 1990s.

Germany became the leading PV market worldwide since revising its Feed-in tariff system as part of the Renewable Energy Sources

Act. Installed PV capacity has risen from 100 MW in 2000 to approximately 4,150 MW at the end of 2007. After 2007, Spain became the largest PV market after adopting a similar feed-in tariff structure in 2004, installing almost half of the photovoltaics (45%) in the world, in 2008, while France, Italy, South Korea and the U.S. have seen rapid growth recently due to various incentive programs and local market conditions. Recent Studies have shown that the global PV market is forecast to exceed 16 GW in the year 2010. The power output of domestic photovoltaic devices is usually described in kilowatt-peak (kWp) units, as most are from 1 to 10 kW.

Development, deployment and economics

The early development of solar technologies starting in the 1860s was driven by an expectation that coal would soon become scarce. However development of solar technologies stagnated in the early 20th century in the face of the increasing availability, economy, and utility of coal and petroleum. The 1973 oil embargo and 1979 energy crisis caused a reorganization of energy policies around the world and brought renewed attention to developing solar technologies. Deployment strategies focused on incentive programs such as the Federal Photovoltaic Utilization Program in the US and the Sunshine Program in Japan. Other efforts included the formation of research facilities in the US (SERI, now NREL), Japan (NEDO), and Germany (Fraunhofer Institute for Solar Energy Systems ISE).

Between 1970 and 1983 photovoltaic installations grew rapidly, but falling oil prices in the early 1980s moderated the growth of PV from 1984 to 1996. Since 1997, PV development has accelerated due to supply issues with oil and natural gas, global warming concerns, and the improving economic position of PV relative to other energy technologies. Photovoltaic production growth has averaged 40% per year since 2000 and installed capacity reached 10.6 GW at the end of 2007, and 14.73 GW in 2008. Since 2006 it has been economical for investors to install photovoltaics for free in return for a long term power purchase agreement. 50% of commercial systems were installed in this manner in 2007 and it is expected that

90% will by 2009. Nellis Air Force Base is receiving photoelectric power for about 2.2 ¢/kWh and grid power for 9 ¢/kWh.

Commercial concentrating solar thermal power (CSP) plants were first developed in the 1980s. CSP plants such as SEGS project in the United States have a levelized energy cost (LEC) of 12-14 ¢/kWh. The 11 MW PS10 power tower in Spain, completed in late 2005, is Europe's first commercial CSP system, and a total capacity of 300 MW is expected to be installed in the same area by 2013.

Operational Solar Thermal Power Stations				
Capacity (MW)	Name	Country	Location	Notes
354	Solar Energy Generating Systems	USA	Mojave Desert California	Collection of 9 units
150	Solnova	Spain	Seville	Completed 2010
100	Andasol solar power station	Spain	Granada	Completed 2009
64	Nevada Solar One	USA	Boulder City, Nevada	
50	Ibersol Ciudad Real	Spain	Puertollano, Ciudad Real	Completed May 2009
50	Alvarado I	Spain	Badajoz	Completed July 2009
50	Extresol 1	Spain	Torre de Miguel Sesmero (Badajoz)	Completed February 2010
50	La Florida	Spain	Alvarado (Badajoz)	completed July 2010

Solar installations in recent years have also largely begun to expand into residential areas, with governments offering incentive programs to make "green" energy a more economically viable option. In Canada the RESOP (Renewable Energy Standard Offer Program), introduced in 2006, and updated in 2009 with the passage of the Green Energy Act, allows residential homeowners in Ontario with solar panel installations to sell the energy they produce back to the grid (i.e., the government) at 42¢/kWh, while drawing power from the grid at an average rate of 6¢/kWh. The program is designed to help promote the government's green agenda and lower the strain often placed on the energy grid at peak hours. In March, 2009 the proposed FIT was increased to 80¢/kWh for small, roof-top systems (≤10 kW).

Photovoltaics are 85 times as efficient as growing corn for ethanol. On a 300 feet (91 m) by 300 feet (1 hectare) plot of land enough ethanol can be produced to drive a car 30,000 miles (48,000 km) per year or 2,500,000 miles (4,020,000 km) by covering the same land with photo cells. The deserts of the South Western United States could produce sufficient electricity to fulfill all of the electrical needs of the United States, and could use electrolysis to produce Hydrogen from water to power aircraft.

World's largest photovoltaic (PV) power plants			
Name of PV power plant	Country	DC Peak Power (MW)	Notes
Sarnia Photovoltaic Power Plant	Canada	80	Completed October 2010
Olmedilla Photovoltaic Park	Spain	60	Completed September 2008
Strasskirchen Solar Park	Germany	54	
Lieberose Photovoltaic Park	Germany	53	2009
Puertollano Photovoltaic Park	Spain	50	2008
Moura photovoltaic power station	Portugal	46	Completed December 2008
Kothen Solar Park	Germany	45	2009
Finsterwalde Solar Park	Germany	42	2009
Waldpolenz Solar Park	Germany	40	550,000 First Solar thin-film CdTe modules. Completed December 2008
Veprek Solar Park	Czech Republic	35.1	186,960 modules, completed September 2010
Planta Solar La Magascona & La Magasquila	Spain	34.5	

The annual International Conference on Solar Photovoltaic Investments, organized by EPIA notes that photovoltaics provides a secure, reliable return on investment, with modules typically lasting 25 to 40 years and with a payback on investment of between 8 to 12 years.

Energy storage methods

Seasonal variation of the output of the solar panels at AT&T Park in San Francisco

Solar energy is not available at night, making energy storage an important issue in order to provide the continuous availability of energy. Both wind power and solar power are intermittent energy sources, meaning that all available output must be taken when it is available and either stored for *when* it can be used, or transported, over transmission lines, to *where* it can be used. Wind power and solar power tend to be somewhat complementary, as there tends to be more wind in the winter and more sun in the summer, but on days with no sun and no wind the difference needs to be made up in some manner. The Institute for Solar Energy Supply Technology of the University of Kassel pilot-tested a combined power plant linking solar, wind, biogas and hydrostorage to provide load-following power around the clock, entirely from renewable sources.

Solar energy can be stored at high temperatures using molten salts. Salts are an effective storage medium because they are low-cost, have a high specific heat capacity and can deliver heat at temperatures compatible with conventional power systems. The Solar Two used this method of energy storage, allowing it to store 1.44 TJ in its 68 m³ storage tank, enough to provide full output for close to 39 hours, with an efficiency of about 99%.

Off-grid PV systems have traditionally used rechargeable batteries to store excess electricity. With grid-tied systems, excess electricity can be sent to the transmission grid. Net metering programs give these systems a credit for the electricity they deliver to the grid. This credit offsets electricity provided from the grid when the system cannot meet demand, effectively using the grid as a storage mechanism. Credits are normally rolled over month to month and any remaining surplus settled annually.

Pumped-storage hydroelectricity stores energy in the form of water pumped when surplus electricity is available, from a lower elevation reservoir to a higher elevation one. The energy is recovered when demand is high by releasing the water: the pump becomes a turbine, and the motor a hydroelectric power generator.

Experimental solar power

Concentrated photovoltaics (CPV) systems employ sunlight concentrated onto photovoltaic surfaces for the purpose of electrical power production. Solar concentrators of all varieties may be used, and these are often mounted on a solar tracker in order to keep the focal point upon the cell as the Sun moves across the sky. Luminescent solar concentrators (when combined with a PV-solar cell) can also be regarded as a CPV system. Luminescent solar concentrators are useful as they can improve performance of PV-solar panels drastically.

Thermoelectric, or "thermovoltaic" devices convert a temperature difference between dissimilar materials into an electric current. First proposed as a method to store solar energy by solar pioneer Mouchout in the 1800s, thermoelectrics reemerged in the Soviet Union during the 1930s. Under the direction of Soviet scientist Abram Ioffe a concentrating system was used to thermoelectrically generate power for a 1 hp engine. Thermogenerators were later used in the US space program as an energy conversion technology for powering deep space missions such as Cassini, Galileo and Viking. Research in this area is focused on raising the efficiency of these devices from 7-8% to 15-20%.

Space-based solar power is a theoretical design for the collection of solar power in space, for use on Earth. SBSP differs from the usual method of solar power collection in that the solar panels used to collect the energy would reside on a satellite in orbit, often referred to as a solar power satellite (SPS), rather than on Earth's surface. In space, collection of the Sun's energy is unaffected by the day/night cycle, weather, seasons, or the filtering effect of Earth's atmospheric gases. Average solar energy per unit area outside Earth's atmosphere is on the order of ten times that available on Earth's surface. However, there is no shortage of energy reaching the surface. The amount of solar energy reaching the surface of the planet each year is about twice the amount of energy that will be obtained forever from coal, oil, natural gas, and mined Uranium, combined, even using breeder reactors.

EPA CO2 EMISSIONS REGULATIONS AND THEIR IMPACT ON INDUSTRIAL CUSTOMERS

Abstract

The United States (U.S.) industrial sector is bracing for new regulations to be issued by the U.S. Environmental Protection Agency to mandate the level of emissions of carbon dioxide (CO2) from manufacturing facilities. These regulations are expected to apply to facilities emitting 25,000 ton of CO2 annually. While the EPA has recently indicated that these rules will not be implemented until the year 2011 given the current economic slowdown, the impacts are expected to be significant. Due to the increased costs associated with CO2 mitigation technologies, the price of all manufactured goods in the economy are anticipated to rise. These increased costs passed through to consumers will accelerate inflation and, in some industries, cause layoffs thereby adding to the current nearly 10% unemployment rate. A recent court ruling in California requires the state to investigate alternative measures, i.e. energy efficiency technologies, compared to mandating the level of emissions of carbon dioxide (CO2) from manufacturing facilities or a cap-and-trade policy.

Discussion

In September, the EPA said it would require coal plants and refineries and other heavy industry facilities emitting more than 25,000 tons a year of greenhouse gases, i.e., CO2, to obtain permits demonstrating they were using the best technology available to reduce emissions blamed for warming the planet.

The agency expects to issue the new rules at the end of May, 2010.

"EPA is considering raising that threshold substantially to reflect input provided during the public comment process," the agency said in a statement.

The EPA was responding to a letter sent recently by Senator Jay Rockefeller, a Democrat from coal-rich West Virginia, and senators from other energy states that expressed concerns about the impact on U.S. workers and business owners of EPA rules that would cut output of the gases from their heavy industry plants.

Sen. Rockefeller said EPA's overture was "good progress."

Nonetheless, he said he would craft legislation to "provide Congress the space it needs" to consider a "workable" climate policy that "will protect jobs and stimulate the economy."

Sources on Capitol Hill were anticipating Rockefeller could introduce a bill imposing a temporary pause, possibly two or three years, for EPA to issue carbon emission-reduction regulations.

"EPA actions in this area would have enormous implications on clean coal state economies and these issues need to be handled carefully and appropriately dealt with by the Congress, not in isolation by a federal environmental agency," Rockefeller said in a statement.

Senator Lisa Murkowski, an Alaska Republican, wants to permanently bar the EPA from regulating greenhouse gas emissions. An aide said she could demand a vote on her bill in the current Congressional year.

EPA administrator Lisa Jackson did not say how much the threshold might be raised. But she expected that EPA would not put any of the new carbon reduction rules in place before 2011.

The agency does not intend to subject smaller plants to permitting any sooner than 2016, she said.

Several states had been concerned EPA was moving too fast. Last month California, the most populous U.S. state, urged the EPA to slow down implementation of the rules, saying they could hurt California's plans to transform its energy system to run more on renewable energy like solar power.

Other states had complained moving quickly could overwhelm their permit offices.

Several environmental groups said the EPA move would create a reasonable timeline to cut emissions. "Just as it has with other pollutants for 40 years, EPA has now made crystal clear that it will address global warming pollution in a way that benefits both our economy and our environment," said Carl Pope, the head of the Sierra Club.

Conclusion

The U.S. industrial sector will likely see climate change regulations, i.e., CO_2 emissions caps, within the next three years. These regulations will either be legislated by the Congress, or at a minimum, be mandated by the EPA. The costs associated with the implementation of these rules will invariably raise the price of finished goods in the economy. Ultimately, the demand for these goods will decline, manufacturing layoffs will occur, unemployment will rise and inflation will increase. A recent court ruling in California requires the state to investigate alternative measures, i.e. energy efficiency technologies, compared to mandating the level of emissions of carbon dioxide (CO_2) from manufacturing facilities or a cap-and-trade policy.

3/31/10

THE NEW NATURAL GAS: DRILLING TECHNOLOGIES AND U.S. PRODUCING ZONES

BY

RICHARD L. ITTEILAG
PRESIDENT
ENERGISTICS, INC.
A DIVISION OF WASHINGTON
PROPOSALASSOCIATES, INC.

Abstract

The United States (U.S.) natural gas resource base is not new. The drilling technologies and resource basins are, while the total resource base continues to be vast and historic. The results of this paper document the new development of shale rock deposits of natural gas and the advanced drilling technologies of horizontal drilling and hydrafracing to produce marketable quantities of natural gas in East Texas, Western Pennsylvania, Southern Arkansas, and Western Louisiana. These shale zones are called: Barnett, Fayetteville, Haynesville and Marcellus.

Discussion

The new natural gas is not a new fossil fuel. It is actually a new set of highly efficient drilling techniques and abundant resource-producing areas in the U.S. These new, resource-rich producing zones are: 1) the Barnett Shale in East Texas, 2) the Marcellus Shale in Western Pennsylvania, 3) the Fayetteville Shale in Arkansas, and 4) the Haynesville Shale in Louisiana. Since these zones are shale rock formations, more sophisticated drilling techniques must be employed then previously utilized in order to extract the natural gas, such as horizontal drilling and hydrafracing.

Of course, the drilling costs per well are higher than traditional wells. For example, in the East Texas Barnett Shale and the Western Pennsylvania Marcellus Shale, the average drilling cost per well is roughly $4.05 per million Btu (mmBtu) according to a recent Wood Mackenzie consultants' study. In the Fayetteville Shale zone in Arkansas, the average drilling costs are slightly higher at about $4.25/ mmBtu. The drilling costs in the Haynesville Shale formations in Louisianna are higher still at about $4.50/mmBtu. These shale natural gas drilling costs are comparable to offshore deep well natural gas drilling costs. With the Obama Administration's recent decision to release offshore producing areas to drilling, the attraction of natural gas shale drilling becomes even more pronounced. Overall, natural gas shale production costs are still much lower and mor economic than producing renewable such as solar energy and wing power.

Conclusion

The United States (U.S.) energy sector has available for current and future production not only traditional lower 48 natural gas reserves, but also significant reserve levels of exotic or unconventional gas reserves in shale rock formations in East Texas, Western Pennsylvania, Southern Arkansas and Western Louisiana. And with significant advances in drilling technologies such as horizontal drilling and hydrafracing, the success rates and drilling costs are currently quite comparable to the economics of offshore drilling and considerably lower than renewables.

8/31/10

UTILITY OFF-PEAK ELECTRICITY LOAD LEVELING PROGRAMS

BY

RICHARD L. ITTEILAG
PRESIDENT
ENERGISTICS, INC.
A DIVISION OF WASHINGTON
PROPOSALASSOCIATES, INC.

Abstract

The resulting economic consequences of a peak electricity shortfall in the United States (U.S.) would be as bad or worse than another oil embargo given the nation's reliance on electricity to cool, heat and power motors and computers. That is the basis of this article: the available energy programs at utilities to reduce electricity peaks or load leveling in the U.S.

Discussion

To date, utility initiatives that encourage cooling, heat and power (CHP) deployment have taken many different forms. Of the 41 U.S. utility companies researched by the Environmental Protection Agency (EPA), 18 provide some type of support for CHP that is not part of a state-mandated initiative; of these, four are gas-only utilities, two are electric and gas utilities, and 12 are electric-only utilities. Eight of the utilities are investor-owned, nine are publicly (municipal- or state-) owned, and one is a cooperative (consumer-owned). Direct financial incentives (i.e., grants or rebates) that are not the result of state policies are not commonly offered. However, at least one investor-owned gas utility (Southwest Gas) and two publicly owned electric utilities (City of Palo Alto Utilities and Sacramento Municipal Utility District) have developed or are developing such programs. Far more common are other types of activities that promote CHP development. The EPA research identified 16 types of utility actions/programs that support CHP development in ways other than offering direct financial incentives for CHP system installation, as follows:

- Program Funding (e.g., system benefits charge) Financial Incentives and State Program Funding
- Request for Proposals (RFP) for Supply CHP Research and Development (R&D)/Demonstration Projects
- Outreach (e.g., CHP-specific Web page)
- Site/Feasibility Analyses
- Design and Engineering
- Construction and Installation
- Maintenance and Operation
- Project Management
- Ownership/Joint Ownership
- Performance Contracting
- Favorable Gas Rates
- Load Curtailment Payments
- Regulatory Process Advice
- Shared Savings Loans (waste heat recovery)
- Custom Rebates (waste heat recovery)

Fuel-Switching Alternatives to Peak Electricity Use

Electricity comprises a 95% market share in the space conditioning segment of the commercial energy market. Natural gas, however, has only a 5% market share of the commercial market for space cooling. This creates a key fuel-switching opportunity to replace peak summertime electricity with off-peak summertime natural gas. In addition, the majority of commercial space cooling equipment is interchangeable between electricity and natural gas. One type of fuel-switching opportunity is:

Ice energy storage, i.e., thermal energy storage (TES), for refrigerant-based air-conditioning systems can be applied to residential and commercial building applications for both new and retrofit facilities. Electric thermal energy storage systems reduce the peak-load energy use by shifting electricity consumption for mechanical refrigeration of water or ice from peak to off-peak hours. The water or ice is kept typically in large insulated storage tanks for use during the next peak usage period.

Conclusion

Essentially, peak electricity demand in the U.S. is reaching crisis or epidemic proportions. A report released by the California Energy Commission in 1996 concluded that it is 8-30 percent more efficient for two major California utilities to produce and transmit a kWh during off-peak hours than during on-peak hours. In response to these findings, California's 2005 release of the Title 24 energy code will surely drive designers to use more efficient, off-peak power because the relative costs are three to four times as high on summer afternoons. The cost of electricity will be priced at the relative cost of energy for every hour of the year (instead of a flat rate as allowed in California's 90.1), otherwise known as 'time dependent valuation.' The similarity here is that through load-leveling programs the problem that a small increase in peak electricity demand could cause a catastrophic crisis in electricity demand across the country could be avoided.

Editor's note-
A Matrix of all the utilities and their programs can be found in:

UTILITY INCENTIVES FOR COMBINED HEAT AND POWER
U. S. Environmental Protection Agency, October 2008
Combined Heat and Power Partnership

5/31/09

ELECTRICITY PEAK-SHAVING TECHNIQUES

BY

RICHARD L. ITTEILAG
PRESIDENT
ENERGISTICS, INC.
A DIVISION OF WASHINGTON
PROPOSALASSOCIATES, INC.

Abstract

While the nation considers the need for energy independence to be critical due to the fact that half the nation's oil consumption is imported, the resulting economic consequences of a peak electricity shortfall would be as bad or worse given the nation's reliance on electricity to cool, light and power motors and computers. That is the nexus of this article: the available energy technologies and programmatic procedures to reduce electricity peaks or peak-shaving in theU.S.

Discussion

As the economy grows at 2-3 percent per year, the total demand for electricity has grown in tandem at 2.1 percent per year over the 1994-2004 period. Inversely, however, electricity capacity margins, the percent of "spinning" supply above demand, have declined consistently over the last decade from 25-30 percent in1992 to about 15 percent today. This compound total demand growth coupled with declines in utility plant capacity margins only masks the serious underlying problem: peak electricity demand, typically for summertime air conditioning, is growing at 2.6 percent per year, consistently as fast as total electricity demand. In fact, the October 16, 2006 report of the North American Electric Reliability Council (NERC) predicts that total United States electricity demand will increase by 19% over the next 10 years while total electricity capacity will rise only 6%. "Capacity margins are projected to drop below minimum target levels in Texas, New England, the Mid-Atlantic area, the Midwest, and the Rocky Mountain area, in the next two to three years, with other portions of the Northeastern U.S., Southwest, and Western U.S. falling below minimum target levels later in the period."

Fuel-Switching Alternatives to Peak Electricity Use

According to the United States Department of Energy (DOE), there are over four million commercial buildings in the United States (US). The largest commercial building sector is the Mercantile and Facilities sector consisting of 22% followed by the Office Building sector comprising 16.4%, Non-Refrigerated Warehouses at 14.6% and Assembly and Education buildings each at 12.6%. Overall, the total market for energy in the commercial sector is fully 33%.

In the space conditioning segment of the commercial energy market, electricity comprises a 95% market share. Natural gas, however, has only a 5% market share of the commercial market for space cooling. This creates a key fuel-switching opportunity to replace peak summertime electricity with off-peak summertime natural gas. In addition, the majority of commercial space cooling equipment is interchangeable between electricity and natural gas. This is evident in the fact that the primary cooling systems utilized in commercial buildings consume electricity or natural gas in the following categories: central cooling systems in 37% of commercial buildings, 31% utilize individual air conditioners, 24% utilize packaged air-conditioning units and 10% use air-source heat pumps. Further, 75% of the commercial buildings cooled utilize duct forced-air distribution systems, while only 11% use fan-coiled units.

Ice energy storage, i.e., thermal energy storage (TES), for refrigerant-based air-conditioning systems can be applied to residential and commercial building applications for both new and retrofit facilities. Electric bills and the attendant electric consumption can be cut significantly by shifting the demand for electricity associated with cooling the air during the peak afternoon times of the day to the off-peak morning and/or nighttime periods of the day to make ice instead. Electric thermal energy storage systems reduce the peak-load energy use by shifting electricity consumption for mechanical refrigeration of water or ice from peak to off-peak hours. The water or ice is kept typically in large insulated storage tanks for use during the next peak usage period.

Both natural gas technologies and thermal energy storage systems exhibit attractive economic scenarios. The simple payback periods are calculated relative to the installed costs of the electric centrifugal system with a hot water boiler In large-scale TES systems due to their inherent economies of scale, cooling applications for sensible heat exhibit a much lower capital cost than latent heat chiller systems, and equally important, a considerably lower capital cost structure than equivalent traditional non-TES electric chiller capacity. The net capital cost savings for TES systems in new construction, in capacity expansion or in capacity rehabilitation is in the magnitude of millions of dollars and, therefore, fairly rapid payback periods.

Energy Policy Act of 2005 (EPACT of 2005)

"Many analysts believe that power prices that vary hour by hour or even minute by minute, abetted by smart meters, are the shape of things to come." The comprehensive array of conservation alternatives that are found in the Energy Policy Act of 2005 (EPACT of 2005) and that create electricity peak-shaving results including smart meters are as follows:

1. Solar (active/passive)
2. Wind
3. Wave

4. Geothermal
5. DDC
6. LEED "green" buildings
7. Demand response
8. Real-time pricing
9. Temperature setbacks
10. Ventilation control
11. Boiler optimization
12. Lighting products/systems
13. Smart meters
14. Reflective roof coatings
15. Cold-water detergents
16. Radiant-heated flooring
17. Concrete construction material
18. LED (light-emitting diodes)lights

Conclusion

Essentially, peak electricity demand in the U.S. is reaching crisis or epidemic proportions. A report released by the California Energy Commission in 1996 concluded that it is 8-30 percent more efficient for two major California utilities to produce and transmit a kWh during off-peak hours than during on-peak hours. The use of more efficient base load generation plants during off-peak hours also lowers transmission and distribution line losses. In addition, cooler nighttime temperatures creates more efficient nighttime generation. In response to these findings, California's 2005 release of the Title 24 energy code will surely drive designers to use more efficient, off-peak power because the relative costs are three to four times as high on summer afternoons. The cost of electricity will be priced at the relative cost of energy for every hour of the year (instead of a flat rate as allowed in California's 90.1), otherwise known as 'time dependent valuation.' The similarity here is the fact that a small increase in peak electricity demand could cause a catastrophic and or an epidemic crisis in electricity demand across the country.

Issues in 21St. Century Energy Economics

Conservation Revisited

Energy efficiency is the 'fifth' energy resource and that, in fact, it is the 'Current' big thing. The United States (U.S.) has made great strides in energy efficiency, i.e., 50% reduction in energy use per unit of output since the 1970s, Europe has lagged far behind. In particular, the Europeans do not place a high priority on energy efficiency. Europe appears to be more concerned with climate change and implementing a 'cap and trade' program to regulate carbon dioxide (CO_2) emissions. While a 'cap and trade' program might have the positive side effects of also reducing energy consumption, i.e., increased conservation, this highly restrictive policy would have the unintended consequences of reducing economic output and exacerbating unemployment when the European economy is at a severely fragile point.

Energy efficiency is like providing the economy a 'fifth' fuel besides natural gas, electricity, oil and coal. Energy efficiency continued to break new ground in the U.S., but at a decreasing pace. Energy efficiency appears to be the best path to improving economic activity, creating new jobs and mitigating climate change.

While it is catchy to say that energy conservation is the 'fifth' energy resource, that would be flat-out wrong! In fact, energy efficiency is the 'faux' energy resource. The United States has been promoting energy efficiency and conservation since the first oil shock in the early 1970s and is still the largest consumer of energy worldwide.

The implications for industry are straight forward. For example, energy conservation can only be effective if the conversion to high-efficiency equipment can pay back the investment in three years or less. Therefore, energy conservation will always compete with other business investments first. Essentially, industry must rely on fossil fuels like natural gas, oil and coal for the next 100 years at a minimum. However, industry must replace obsolete rquipment with new, high-efficiency equipment over the useful life of the equipment. This will ensure that the energy footprint for industry will contract over time, but at a very slow, deliberate pace. The only conservation and renewable fuel that the U.S. can rely on today is hydro power, but only for less than 5% of electric power generation.

For the immediate timeframe, industry will be faced with mandated efficiency standards in the form of 'demand response' programs. These required efficiency standards are prominent in the western U.S., primarily in California, but are rapidly being adopted nationwide. Ultimately, from an industry perspective, energy efficiency or conservation can reduce industry's input costs and achieve higher product margins while being good corporate citizens and silent environmentalists.

Energy Independence and Policy Act of 2007 (EPact of 2007)

Lighting Efficiency

1. appliance bulbs, "rough service" bulbs, colored bulbs, plant lights and 3-way bulbs, are exempt from 25% greater efficiency standards
2. 40 watt and more than 150 watt-plus bulbs are also exempt
3. stage and Malibu lighting are exempted, as well

Buildings Efficiency

1. creates Office of Commercial High Performance Green Buildings in the Department of Energy (DOE) to promote more efficient buildings
2. creates a nationwide zero-net energy initiative for commercial buildings built after 2025

Industrial Efficiency

1. DOE must research and develop ways to improve the energy efficiency of equipment and industrial processes

2. the Environmental Protection Agency (EPA) must create a waste energy recovery program

Institutional Establishments

1. creates grants to support improved energy efficiency and sustainability at public institutions

Tax Deductions

1. $1.80/square foot for energy expenses that improve building efficiency by 50% compared to 2001 use patterns (ASHRAE 90.1)
2. system-specific deductions up to $0.60/sq. ft.
   a. interior lighting systems that are substantially lower than standard lighting requirements
   b. $0.60/sq. ft. for 40% below standard and $0.30/sq.ft. for 25% below standard
   c. a sliding scale deduction between $0.30/sq.ft. to $0.60/sq. ft. for lighting systems that are 25%-40% more efficient

Proposed EPA Carbon Dioxide (CO2) Standard for New Power Plants

Consistent with the U. S. Supreme Court's decision in 2009, the Environmental Protection Agency (EPA) determined that greenhouse gas (CO2) pollution threatens Americans' health and welfare by leading to long lasting changes in our climate. These changes can have a range of negative effects on human health and the environment.

On March 27, 2012, EPA proposed a CO2 pollution standard on the construction of new power plants. The new standard does not apply to plants currently operating or new permitted plants that begin construction over the next 12 months. The standard would be flexible and would minimize CO2 pollution through the deployment of the same types of modern technologies nationwide currently employed in new power plant construction today.

For the purpose of this standard, the proposed rule would apply only to NEW fossil fuel-fired electric generating units (EGUs) larger than 25 megawatts (MW). New plants can choose to burn any fossil fuel to generate electricity for sale, including natural gas as well as coal with the appropriate technologies that reduce CO_2 emissions.

The EPA is proposing that new fossil fuel power plants meet an output-based standard of 1,000 pounds of CO_2 per megawatt hour (lb CO_2/MWh gross). Clearly, natural gas power plants could meet this standard without any high-cost plant pollution equipment. Coal power plants, however, would be essentially prohibited due to the requirement for new CO_2 pollution equipment.

However, a new coal plant could still be built if it employed technology to reduce CO_2 emissions to meet the standard, such as carbon capture and storage (CCS). New coal plants employing CCS would have the option to use a 30-year average of CO_2 emissions to meet the proposed standard, rather than meeting the annual standard each year.

   a. coal plants that install and operate CCS immediately would have the flexibility to emit more CO_2 in the early years, as they learn how to best optimize the controls.
   b. a utility could build a coal power plant and add CCS later, I.e., a new coal plant could be built and emit more CO_2 for the first 10 years then emit less CO_2 for the next 20 years, as long as the average of those emissions meet the standard.
   c. CSS technology is expected to become more widely available, which should lead to lower costs and improved performance over time.

EPA's proposed standard reflects the ongoing trend in the power sector to build cleaner plants, including new, clean burning, efficient natural gas generation, which is already the fuel of choice for most new and planned power plants.

At the same time, the standard creates a path forward for new technologies to be developed at future facilities that would allow utilities to burn coal, while emitting less CO2 pollution.

New natural gas combined cycle (NGCC) power plant units should be able to meet the proposed standard without add-on controls. In fact, EPA estimates that 95% of the NGCC plants built since 2005 would meet the standard.

New power plants that are designed to burn coal would be able to incorporate technology to reduce CO2 emissions to meet the standard, such as CCS. Nationwide, states like Washington, Oregon and California, currently limit CO2 emissions. Other states, like Montana and Illinois, currently require CCS for new coal-fired power plants.

EPA also believes the standard allows flexibility to allow utilities to phase-in over tine technology to reduce CO2. And EPA suggests that the standard is fully compatible with current utility industry investment patterns resulting in cost containment for compliance.

Thus, while it will be highly restrictive, expensive and time consuming due to the permitting process to build new coal plants in the future, it will not be entirely prohibited by the imposition of this EPA standard.

Climate Science

1. 650 million years ago, earth, or rodinia, was covered in ice
2. the Cambrian Explosion warmed the earth and created its atmosphere, therefore, life began 500 million years ago
3. the Cambrian Seas brought forth modern animals on earth and beneath the sea
4. the ozone layer was formed and the earth's atmosphere was created
5. 400 million years ago the earth was born or 4 billion years after the earth was formed

6. coal was formed 200 million years ago from decaying plant life
7. 300 million years ago oil and gas were formed from fossils and the earth was 4.5 billion years old
8. the Great Pangea occurred 250 million years ago with an entire earth volcanic eruption annihilating all life
9. dinosaur means terrible lizard
10. 150 million years ago the earth's current continents were formed and the earth had its first "global warming"
11. dinosaurs ruled earth 100 million years ago and produced earth's diamonds
12. 1869 diamonds found in Kimberley area of S. Africa in volcanic mines
13. Pangea split apart 100 million years ago
14. 50-65 million years ago dinosaurs disappeared
15. iridium found in space rock on earth
16. 1960 the Alvarez theory of an asteroid/meteor event killing the dinosaurs was introduced
17. 65 million years ago there was a giant meteor/asteroid strike on earth destroying the dinosaurs
18. the collision of African & European continents form the Alps, I.e., the Materhorn
19. water erosion causes the height of mountains
20. Grand Canyon formed 6 million years ago and was formed by plate tectonics
21. earth ice ages formed
22. last ice age 10,000 years ago (global warming)
23. 4.5-5 billion years ago the earth was formed and has fluctuated constantly meteorologically
24. earth is in a brief warm period between ice ages expected about 15,000 years from today
25. earth will end billions of years from now due to the end of plate tectonics

Global Warming Debate

There appears, as 2011 comes to an end, that there is a lack of "absolute" proof of global warming. Bjorn Lomborg, author of "Cool

It" and adjunct professor at Copenhagen Business School, suggests in an Op Ed page article in the Wall St. Journal on December 12, 2011, "Global Warming and Adaptability," we look for 'adaptive' steps to deal with environmental changes as whether man's actions are the cause or not. Any carbon dioxide (CO2) emissions reduction deal internationally by the United Nations to replace Kyoto would only have a 'negligible' impact on climate change, I.e., global warming, in coming decades. According to Lomborg, an expert in the climate debate, in order to help real people we need to focus first on 'adaptability.' Even if the world were to cut carbon emissions by 50% below 1990-levels by 2050, which is highly unlikely, the difference in temperature would be less than 0.2 degrees Fahrenheit in 2050! The world would be better served by improving crop-yield to feed the world's starving population.

In addition, Professor Henrik Svensmark, who is at the Danish National Space Institute in Copenhagen, has observed that sun spots help to control cloud cover on the earth which absorbs energy in the atmosphere from the sun causing warming and cooling cycles. This mechanism, if proven definitively, has little do do with carbon emissions in our atmosphere, and is not man-made nor controllable by man.

The real problem is the science. What makes a greenhouse gas function in the role of warming the atmosphere is the ability to absorb infrared radiation, and this depends on the molecular vibrations of the molecule that allow the molecules to absorb and re-emit incident radiation. Carbon Dioxide (GO2 weight 44), is heavier than nitrogen (NI weight 28) or oxygen (O weight 32) and those are the major gaseous components in the atmosphere, but the lighter gases, water (H2O weight 18) or methane (CH4 weight 16), also found in the atmosphere, are much stronger infrared absorbers by virtue of their OH and CH chemical bonds, respectively. Methane is increasing slowly but carbon dioxide is still in greater concentration in the atmosphere and is more under control of mankind than methane which results mainly from anaerobic decomposition of organic matter. Therefore, adaptability is where our focus should be, not on a 'cap and trade' tax increase policy during a deep recession in the U.S. and Europe.

Future Competitive Power Generation Options

There have been many studies carried out examining the economics of various future generation options, and the following are merely the most important and also focus on the nuclear element.

A 2010 OECD study Projected Costs of Generating Electricity compared 2009 data for generating base-load electricity by 2015 as well as costs of power from renewables, and showed that nuclear power was very competitive at $30 per ton CO_2 cost and low discount rate. The study comprised data for 190 power plants from 17 OECD countries as well as some data from Brazil, China, Russia and South Africa. It used levelised lifetime costs with carbon price internalised (OECD only) and discounted cash flow at 5% and 10%. The precise competitiveness of different base-load technologies depended very much on local circumstances and the costs of financing and fuels.

Nuclear overnight capital costs in the OECD ranged from US$ 1556/kW for APR-1400 in South Korea through $3009 for ABWR in Japan, $3382/kW for Gen III+ in USA, $3860 for EPR at Flamanville in France to $5863/kW for EPR in Switzerland, with world median $4100/kW. Belgium, Netherlands, Czech Rep and Hungary were all over $5000/kW. In China overnight costs were $1748/kW for CPR-1000 and $2302/kW for AP1000, and in Russia $2933/kW for VVER-1150. EPRI (USA) gave $2970/kW for APWR or ABWR, Eurelectric gave $4724/kW for EPR. OECD black coal plants were costed at $807-2719/kW, those with carbon capture and compression (tabulated as CCS, but the cost not including storage) at $3223-5811/kW, brown coal $1802-3485, gas plants $635-1747/kW and onshore wind capacity $1821-3716/kW. (Overnight costs were defined here as EPC, owner's costs and contingency, but excluding interest during construction.)

OECD electricity generating cost projections for year 2010 on -5% discount rate, c/kWh					
Country	Nuclear	Coal	coal with CCS	Gas CCGT	Onshore wind
Belgium	6.1	8.2	-	9.0	9.6
Czech R	7.0	8.5-9.4	8.8-9.3	9.2	14.6
Germany	5.0	7.0-7.9	6.8-8.5	8.5	10.6
Hungary	8.2	-	-	-	-
Japan	5.0	8.8	-	10.5	-
Korea	2.9-3.3	6.6-6.8	-	9.1	-
Netherlands	6.3	8.2	-	7.8	8.6
Slovakia	6.3	12.0	-	-	-
Switzerland	5.5-7.8	-	-	9.4	16.3
USA	4.9	7.2-7.5	6.8	7.7	4.8
China*	3.0-3.6	5.5	-	4.9	5.1-8.9
Russia*	4.3	7.5	8.7	7.1	6.3
EPRI (USA)	4.8	7.2	-	7.9	6.2
Eurelectric	6.0	6.3-7.4	7.5	8.6	11.3

* For China and Russia: 2.5c is added to coal and 1.3c to gas as carbon emission cost to enable sensible comparison with other data in those fuel/technology categories, though within those countries coal and gas will in fact be cheaper than the Table above suggests.

Source: OECD/IEA NEA 2010, table 4.1.

At 5% discount rate comparative costs are as shown above. Nuclear is comfortably cheaper than coal and gas in all countries. At 10% discount rate (below) nuclear is still cheaper than coal in all but the Eurelectric estimate and three EU countries, but in these three gas becomes cheaper still. Coal with carbon capture is mostly more expensive than either nuclear or paying the $30 per tonne for CO2 emissions, though the report points out "great uncertainties" in the cost of projected CCS. Also, investment cost becomes a much greater proportion of power cost than with 5% discount rate.

OECD electricity generating cost projections for year 2010 on -10% discount rate, c/kWh					
Country	Nuclear	Coal	Coal with CCS	Gas CCGT	Onshore wind
Belgium	10.9	10.0	-	9.3-9.9	13.6
Czech R	11.5	11.4-13.3	13.6-14.1	10.4	21.9
France	9.2	-	-	-	12.2
Germany	8.3	8.7-9.4	9.5-11.0	9.3	14.3
Hungary	12.2	-	-	-	-
Japan	7.6	10.7	-	12.0	-
Korea	4.2-4.8	7.1-7.4	-	9.5	-
Netherlands	10.5	10.0	-	8.2	12.2
Slovakia	9.8	14.2	-	-	-
Switzerland	9.0-13.6	-	-	10.5	23.4
USA 7.7	8.8-9.3	9.4	8.3	7.0	5.1-8.9
China*	4.4-5.5	5.8	-	5.2	7.2-12.6
Russia*	6.8	9.0	11.8	7.8	9.0
EPRI (USA)	7.3	8.8-8.3	9.1	8.6	11.3
Eurelectric	10.6	8.0-9.0	10.2	9.4	15.5

* For China and Russia: 2.5c is added to coal and 1.3c to gas as carbon emission cost to enable sensible comparison with other data in those fuel/technology categories, though within those countries coal and gas will in fact be cheaper than the Table above suggests.

Source: OECD/IEA NEA 2010, table 4.1.

A 2004 report from the University of Chicago, funded by the US Department of Energy, compared the levelised power costs of future nuclear, coal, and gas-fired power generation in the USA. Various nuclear options were covered, and for an initial ABWR or AP1000 they range from 4.3 to 5.0 c/kWh on the basis of overnight capital costs of $1200 to $1500/kW, 60 year plant life, 5 year construction and 90% capacity. Coal gives 3.5-4.1 c/kWh and gas (CCGT) 3.5-4.5 c/kWh, depending greatly on fuel price.

The levelised nuclear power cost figures include up to 29% of the overnight capital cost as interest, and the report notes that up to another 24% of the overnight capital cost needs to be added for the initial unit of a first-of-a-kind advanced design such as the AP1000, defining the high end of the range above. For more advanced plants such as the EPR or SWR1000, overnight capital cost of $1800/kW is assumed and power costs are projected beyond the range above. However, considering a series of eight units of the same kind and assuming increased efficiency due to experience which lowers overnight capital cost, the levelised power costs drop 20% from those quoted above and where first-of-a-kind engineering costs are amortised (eg the $1500/kW case above), they drop 32%, making them competitive at about 3.4 c/kWh.

Nuclear Plant: Projected Electricity Prices (c/kWh)			
Overnight capital cost $/kW	1200	1500	1800
First unit	7 yr build, 40 yr life		
5.3			
6.2			
7.1			
	5 yr build, 60 yr life		
4.3			
5.0			
5.8			
4th unit	7 yr build, 40 yr life		
4.5			
4.5			
5.3			
	5 yr build, 60 yr life *		
3.7			
3.7			
4.3			
8th unit	7 yr build, 40 yr life		
4.2			
4.2			
4.9			
	5 yr build, 60 yr life *		
3.4			
3.4			
4.0			

* calculated from above data

The study also shows that with a minimal carbon control cost impact of 1.5 c/kWh for coal and 1.0 c/kWh for gas superimposed on the above figures, nuclear is even more competitive. But more importantly it goes on to explore other policy options which would offset investment risks and compensate for first-of-a-kind engineering costs to encourage new nuclear investment, including investment tax breaks, and production tax credits phasing out after 8 years. (US wind energy gets a production tax credit which has risen to 2.1 c/kWh.)

In May 2009 an update of a heavily-referenced 2003 MIT study was published. This said that "since 2003 construction costs for all types of large-scale engineered projects have escalated dramatically. The estimated cost of constructing a nuclear power plant has increased at a rate of 15% per year heading into the current economic downturn. This is based both on the cost of actual builds in Japan and Korea and on the projected cost of new plants planned for in the United States. Capital costs for both coal and natural gas have increased as well, although not by as much. The cost of natural gas and coal that peaked sharply is now receding. Taken together, these escalating costs leave the situation [of relative costs] close to where it was in 2003." The overnight capital cost was given as $4000/kW, in 2007 dollars. Applying the same cost of capital to nuclear as to coal and gas, nuclear came out at 6.6 c/kWh, coal at 8.3 cents and gas at 7.4 cents, assuming a charge of $25/tonne CO_2 on the latter.

Escalating capital costs were also highlighted in the US Energy Information Administration (EIA) 2010 report "Updated Capital Cost Estimates for Electricity Generation Plants". The US cost estimate for new nuclear was revised upwards from $3902/kW by 37% to a value of $5339/kW for 2011 by the EIA. This is in contrast to coal, which increases by only 25%, and gas which actually shows a 3% decrease in cost. Renewables estimates show solar dropping by 25% while onshore wind increases by about 21%. The only option to increase faster than nuclear is offshore wind at 49%, while the increase in coal with CCS is about the same as nuclear. In the previous year's estimate, EIA assumed that the cost of nuclear would drop with time and experience, and that by 2030 the cost of nuclear would drop by almost 30% in constant dollars.

By way of contrast, China is stating that it expects its costs for plants under construction to come in at less than $2000/kW and that subsequent units should be in the range of $1600/kW. This estimates is for the AP1000 design, the same as used by EIA for the USA. This would mean that an AP1000 in the USA would cost about three times as much as the same plant built in China. Different labour rates in the two countries are only part of the explanation. Standardised design, numerous units being built, and increased localization are all significant factors in China.

The French Energy & Climate Directorate published in November 2008 an update of its earlier regular studies on relative electricity generating costs. This shied away from cash figures to a large extent due to rapid changes in both fuel and capital, but showed that at anything over 6000 hours production per year (68% capacity factor), nuclear was cheaper than coal or gas combined cycle (CCG). At 100% capacity CCG was 25% more expensive than nuclear. At less than 4700 hours per year CCG was cheapest, all without taking CO2 cost into account.

With the nuclear plant fixed costs were almost 75% of the total, with CCG they were less than 25% including allowance for CO2 at $20/t. Other assumptions were 8% discount rate, gas at 6.85 $/GJ, coal at EUR 60/t. The reference nuclear unit is the EPR of 1630 MWe net, sited on the coast, assuming all development costs being borne by Flamanville 3, coming on line in 2020 and operating only 40 of its planned 60 years. Capital cost apparently EUR 2000/kW. Capacity factor 91%, fuel enrichment is 5%, burnup 60 GWd/t and used fuel is reprocessed with MOX recycle. In looking at overall fuel cost, uranium at $52/lb made up about 45% of it, and even though 3% discount rate was used for back-end the study confirmed the very low cost of waste in the total—about 13% of fuel cost, mostly for reprocessing.

At the end of 2008 EdF updated the overnight cost estimate for Flamanville 3 EPR (the first French EPR, but with some supply contracts locked in before escalation) to EUR 4 billion in 2008 Euros (EUR 2434/kW), and electricity cost 5.4 cents/kWh (compared with 6.8 c/kWh for CCGT and 7.0 c/kWh for coal, "with lowest assumptions" for CO_2 cost). These costs were confirmed in mid 2009, when EdF had spent nearly EUR 2 billion. In July 2010 EdF revised the overnight cost to about EUR 5 billion.

A detailed study of energy economics in Finland published in mid 2000 was important in making the strong case for additional nuclear construction there, showing that nuclear energy would be the least-cost option for new generating capacity. The study compared nuclear, coal, gas turbine combined cycle and peat. Nuclear has very much higher capital costs than the others—EUR 1749/kW including initial fuel load, which is about three times the cost of the gas plant. But its fuel costs are much lower, and so at capacity factors above 64% it is the cheapest option.

August 2003 figures put nuclear costs at EUR 2.37 c/kWh, coal 2.81 c/kWh and natural gas at 3.23 c/kWh (on the basis of 91% capacity factor, 5% interest rate, 40 year plant life). With emission trading @ EUR 20/t CO_2, the electricity prices for coal and gas increase to 4.43 and 3.92 c/kWh respectively:

In the middle three bars of this graph the relative effects of capital and fuel costs can be clearly seen. The relatively high capital cost of nuclear power means that financing cost and time taken in construction are critical, relative to gas and even coal. But the fuel cost is very much lower, and so once a plant is built its cost of production is very much more predictable than for gas or even coal. The impact of adding a cost or carbon emissions can also be seen.

There have been a large number of recent estimates from the United States of the costs of new nuclear power plants. For example, Florida Power & Light in February 2008 released projected figures for two new AP1000 reactors at its proposed Turkey Point site. These took into account increases of some 50% in material, equipment and labour since 2004. The new figures for overnight capital cost ranged from $2444 to $3582 /kW, or when grossed up to include cooling towers, site works, land costs, transmission costs and risk management, the total cost came to $3108 to $4540 per kilowatt. Adding in finance charges almost doubled the overall figures at $5780 to $8071 /kW. FPL said that alternatives to nuclear for the plant were not economically attractive.

In May 2008 South Carolina Electric and Gas Co. and Santee Cooper locked in the price and schedule of new reactors for their Summer plant in South Carolina at $9.8 billion. (The budgeted cost earlier in the process was $10.8 billion, but some construction and material costs ended up less than projected.) The EPC contract for completing two 1,117-MW AP1000s is with Westinghouse and the Shaw Group. Beyond the cost of the actual plants, the figure includes forecast inflation and owners' costs for site preparation, contingencies and project financing. The units are expected to be in commercial operation in 2016 and 2019.

In November 2008 Duke Energy Carolinas raised the cost estimate for its Lee plant (2 x 1117 MWe AP1000) to $11 billion, excluding finance and inflation, but apparently including other owners costs.

In November 2008 TVA updated its estimates for Bellefonte units 3 & 4 for which it had submitted a COL application for twin AP1000 reactors, total 2234 MWe. It said that overnight capital cost estimates ranged from $2516 to $4649/kW for a combined construction cost of $5.6 to 10.4 billion. Total cost to the owners would be $9.9 to $17.5 billion.

Regarding bare plant costs, some recent figures apparently for overnight capital cost (or Engineering, Procurement and Construction—EPC—cost) quoted from reputable sources but not necessarily comparable are:

- EdF Flamanville EPR: EUR 4 billion/$5.6 billion, so EUR 2434/kW or $3400/kW

- Bruce Power Alberta 2x1100 MWe ACR, $6.2 billion, so $2800/kW
- CGNPC Hongyanhe 4x1080 CPR-1000 $6.6 billion, so $1530/kW
- AEO Novovronezh 6&7 2136 MWe net for $5 billion, so $2340/kW
- AEP Volgodonsk 3 & 4, 2 x 1200 MWe VVER $4.8 billion, so $2000/kW
- KHNP Shin Kori 3&4 1350 MWe APR-1400 for $5 billion, so $1850/kW
- FPL Turkey Point 2 x 1100 MWe AP1000 $2444 to $3582/kW
- Progress Energy Levy county 2 x 1105 MWe AP1000 $3462/kW
- NRG South Texas 2 x 1350 MWe ABWR $8 billion, so $2900/kW
- ENEC for UAE from Kepco, 4 x 1400 MWe APR-1400 $20.4 billion, so $3643/kW

A striking indication of the impact of financing costs is given by Georgia Power, which said in mid 2008 that twin 1100 MWe AP1000 reactors would cost $9.6 billion if they could be financed progressively by ratepayers, or $14 billion if not. This gives $4363 or $6360 per kilowatt including all other owners costs.

Finally, in the USA the question of whether a project is subject to regulated cost recovery or is a merchant plant is relevant, since it introduces political, financial and tactical factors. If the new build cost escalates (or is inflated), some cost recovery may be possible through higher rates can be charged by the utility if those costs are deemed prudent by the relevant regulator. By way of contrast, a merchant plant has to sell all its power competitively, so must convince its shareholders that it has a good economic case for moving forward with a new nuclear unit.

External costs

The report of a major European study of the external costs of various fuel cycles, focusing on coal and nuclear, was released in mid 2001—ExternE. It shows that in clear cash terms nuclear energy incurs about one tenth of the costs of coal. The external costs are defined as those actually incurred in relation to health and the environment and quantifiable but not built into the cost of the electricity. If these costs were in fact included, the EU price of electricity from coal would double and that from gas would increase 30%. These are without attempting to include the external costs of global warming.

The European Commission launched the project in 1991 in collaboration with the US Department of Energy, and it was the first research project of its kind "to put plausible financial figures against damage resulting from different forms of electricity production for the entire EU". The methodology considers emissions, dispersion and ultimate impact. With nuclear energy the risk of accidents is factored in along with high estimates of radiological impacts from mine tailings (waste management and decommissioning being already within the cost to the consumer). Nuclear energy averages 0.4 euro cents/kWh, much the same as hydro, coal is over 4.0 cents (4.1-7.3), gas ranges 1.3-2.3 cents and only wind shows up better than nuclear, at 0.1-0.2 cents/kWh average. NB these are the external costs only.

Sources:

OECD/ IEA NEA 2010, Projected Costs of Generating Electricity.
OECD, 1994, The Economics of the Nuclear Fuel Cycle.
NEI: US generating cost data.
Tarjanne, R & Rissanen, S, 2000, Nuclear Power: Lest-cost option for baseload electricity in Finland; in Proceedings 25th International Symposium, Uranium Institute.
Gutierrez, J 2003, Nuclear Fuel—key for the competitiveness of nuclear energy in Spain, WNA Symposium.

University of Chicago, August 2004, The Economic Future of Nuclear Power.
Nuclear Energy Institute, August 2008, The cost of new generating capacity in perspective.

Peak Electricity Capacity
in US Is Supply
Constrained
Nuclear Power Plants
Are Prohibited from
Populated Areas
10-15 Year Construction Lead-Times for Nuclear
Power Plants
Strict State Reliability Standards Enforced Nationwide for Nuclear Power Plants
Nuclear Plant Capital Costs Are 25%-37% More Expensive Than Natural Gas
Coal Plant CO_2 Restrictions due to Clean Air Act Provisions
Vast Natural Gas
Pipeline Network
Throughout U.S.
Ample Natural Gas Supply
in Shale Formations Domestically (TX, LA, PA, NY)
Nuclear Waste Storage and
Transportation Are Problematic (Environmentally, National Security, Safety, etc,)

Electricity Capacity Enhancements

1. high-efficiency equipment
2. shifting load off peak
3. smart metering
4. time-of-use pricing
5. renewable resources (solar, wind, geothermal, tidal)

"Capital Cost Advantage:
Natural Gas v. Nuclear Electric Generation"

While nuclear electric generation has been tauted by the nuclear industry as the panacea for building new power generation capacity, desperately needed in the United States today, the economics do not bear that opinion out. To begin with, nuclear plant capital costs are some 25%-37% more expensive than comparable natural gas-fired power generation.

In addition, from a purely environmental perspective, natural gas produces 28 watts of electricity per square meter of land used to drill a standard natural gas well. While it is true that nuclear plant power generation produces a larger amount of watts per square meter, I. e., about 2,000 watts, this is dwarfed by the significant capital cost disadvantage.

The first expansion of a nuclear power plant since the 1970s was approved for construction recently in Georgia by the Southern Company for expected completion in 2014. The 100-plus nuclear plants operating in the U. S. currently provide 20% of the U.S. electricity capacity. Cost overruns and safety concerns following the 1979 Three Mile Island accident stalled new construction. In fact, even this recent approval in Georgia of a permit by the Nuclear Regulatory Commission (NRC) was not without controversy. The NRC Chairman, Gregory Jaczko, opposed the Vogtle plant. He proposed that a condition be included in the permit that before operating the two reactors, Southern should follow any forthcoming safety enhancements put forward by the NRC as it reviews the meltdown last year at Japan's Fukushima Daiichi nuclear power plant. In addition, Southern needs a federal loan guarantee to help finance the $14 billion project, which it will operate and co-own with three other partners.

The outlook for additional nuclear plants in the near term isn't promising either because of low electricity demand resulting from the deep recession and the sharp decline in natural gas prices, the main competing fuel to nuclear power generation. The Vogtle reators and a pair of South Carolina reactors will surely be the only facilities built in the U.S. before 2020.

In fact, existing nuclear facilities are facing questions on their safety. The San Onofre Nuclear Generating Station in southern California has been shut down due to deteriorating steam pipes in both reactors. This development has heightened public concern in California, specifically, but it is echoed nationwide, to permanently close the plant cutting-off 2,200 megawatts of generating capacity in an already tight electricity capacity environment. This situation will only worsen in the near-term future as nuclear plants are retired and not replaced.

Nuclear power is cost competitive with other forms of electricity generation, except where there is direct access to low-cost fossil fuels.

Fuel costs for nuclear plants are a minor proportion of total generating costs, though capital costs are greater than those for coal-fired plants and much greater than those for gas-fired plants.

In assessing the economics of nuclear power, decommissioning and waste disposal costs are fully taken into account.

Assessing the relative costs of new generating plants utilising different technologies is a complex matter and the results depend crucially on location. Coal is, and will probably remain, economically attractive in countries such as China, the USA and Australia with abundant and accessible domestic coal resources as long as carbon emissions are cost-free. Gas is also competitive for base-load power in many places, particularly using combined-cycle plants, especially ad gas prices have remained low giving gas generation a competitive advantage.

Assessing the relative costs of new generating plants utilising different technologies is a complex matter and the results depend crucially on location. Coal is, and will probably remain, economically attractive in countries such as China, the USA and Australia with abundant and accessible domestic coal resources as long as carbon emissions are cost-free. Gas is also competitive for base-load power in many places, particularly using combined-cycle plants, though rising gas prices have removed much of the advantage.

Nuclear energy is, in many places, particularly the developing world, competitive with fossil fuels for electricity generation, despite relatively high capital costs and the need to internalise all waste disposal and decommissioning costs. If the social, health and environmental costs of fossil fuels are also taken into account, the economics of nuclear power are suspect, at best.

The Economics of Nuclear Power

Nuclear power is cost competitive with other forms of electricity generation, except where there is direct access to low-cost fossil fuels.

Fuel costs for nuclear plants are a minor proportion of total generating costs, though capital costs are greater than those for coal-fired plants and much greater than those for gas-fired plants.

In assessing the economics of nuclear power, decommissioning and waste disposal costs are fully taken into account.

Assessing the relative costs of new generating plants utilising different technologies is a complex matter and the results depend crucially on location. Coal is, and will probably remain, economically attractive in countries such as China, the USA and Australia with abundant and accessible domestic coal resources as long as carbon emissions are cost-free. Gas is also competitive for base-load power in many places, particularly using combined-cycle plants, because gas prices have remained low.

Nuclear energy is, in many places, competitive with fossil fuels for electricity generation, despite relatively high capital costs and the need to internalise all waste disposal and decommissioning costs. If the social, health and environmental costs of fossil fuels are also taken into account, the economics of nuclear power are outstanding.

The New Economics of Nuclear Power

From the outset the basic attraction of nuclear energy has been its low fuel costs compared with coal, oil and gas-fired plants. Uranium, however, has to be processed, enriched and fabricated into fuel elements, and about half of the cost is due to enrichment and fabrication. In the assessment of the economics of nuclear power allowances must also be made for the management of radioactive used fuel and the ultimate disposal of this used fuel or the wastes separated from it. But even with these included, the total fuel costs of a nuclear power plant in the OECD are typically about a third of those for a coal-fired plant and between a quarter and a fifth of those for a gas combined-cycle plant. The US Nuclear Energy Institute suggests that for a coal-fired plant 78% of the cost is the fuel, for a gas-fired plant the figure is 89%, and for nuclear the uranium is about 14%, or double that to include all front end costs.

In March 2011, the approx. US $ cost to get 1 kg of uranium as UO2 reactor fuel (at current spot uranium price):

Uranium:	8.9 kg U3O8 x $146	US$ 1300
Conversion:	7.5 kg U x $13	US$ 98
Enrichment:	7.3 SWU x $155	US$ 1132
Fuel fabrication:	per kg	US$ 240
Total, approx:		US$ 2770

At 45,000 MWd/t burn-up this gives 360,000 kWh electrical per kg, hence fuel cost: 0.77 c/kWh.

Fuel costs are one area of steadily increasing efficiency and cost reduction. For instance, in Spain the nuclear electricity cost was reduced by 29% over 1995-2001. This involved boosting enrichment levels and burn-up to achieve 40% fuel cost reduction. Prospectively, a further 8% increase in burn-up will give another 5% reduction in fuel cost.

Uranium has the advantage of being a highly concentrated source of energy which is easily and cheaply transportable. The quantities needed are very much less than for coal or oil. One kilogram of natural uranium will yield about 20,000 times as much energy as the same amount of coal. It is therefore intrinsically a very portable and tradeable commodity.

The fuel's contribution to the overall cost of the electricity produced is relatively small, so even a large fuel price escalation will have relatively little effect (see below). Uranium is abundant.

There are other possible savings. For example, if used fuel is reprocessed and the recovered plutonium and uranium is used in mixed oxide (MOX) fuel, more energy can be extracted. The costs of achieving this are large, but are offset by MOX fuel not needing enrichment and particularly by the smaller amount of high-level wastes produced at the end. Seven UO2 fuel assemblies give rise to one MOX assembly plus some vitrified high-level waste, resulting in only about 35% of the volume, mass and cost of disposal.

Comparing the Economics of Electricity Generation

It is important to distinguish between the economics of nuclear plants already in operation and those at the planning stage. Once capital investment costs re effectively "sunk", existing plants operate at very low costs and are effectively "cash machines". Their operations and maintenance (O&M) and fuel costs (including used fuel management) are, along with hydropower plants, at the low end of the spectrum and make them very suitable as base-load power suppliers. This is irrespective of whether the investment costs are amortized or depreciated in corporate financial accounts—assuming the forward or marginal costs of operation are below the power price, the plant will operate.

US figures for 2008 published by NEI show the general picture, with nuclear generating power at 1.87 c/kW.o

Note: the above data refer to fuel plus operation and maintenance costs only, they exclude capital, since this varies greatly among utilities and states, as well as with the age of the plant. A Finnish study in 2000 also quantified fuel price sensitivity to electricity costs:

These show that a doubling of fuel prices would result in the electricity cost for nuclear rising about 9%, for coal rising 31% and for gas 66%. Gas prices have since risen significantly.

The impact of varying the uranium price in isolation is shown below in a worked example of a typical US plant, assuming no alteration in the tails assay at the enrichment plant.

Doubling the uranium price (say from $25 to $50 per lb U3O8) takes the fuel cost up from 0.50 to 0.62 US cents per kWh, an increase of one quarter, and the expected cost of generation of the best US plants from 1.3 US cents per kWh to 1.42 cents per kWh (an increase of almost 10%). So while there is some impact, it is comparatively minor, especially by comparison with the impact of gas prices on the economics of gas generating plants. In these, 90% of the marginal costs can be fuel. Only if uranium prices rise to above $100 per lb U3O8 ($260 /kgU) and stay there for a prolonged period (which seems very unlikely) will the impact on nuclear generating costs be considerable.

Nevertheless, for nuclear power plants operating in competitive power markets where it is impossible to pass on any fuel price increases (ie the utility is a price-taker), higher uranium prices will cut corporate profitability. Yet fuel costs have been relatively stable over time—the rise in the world uranium price between 2003 and 2007 added to generation costs, but conversion, enrichment and fuel fabrication costs did not followed the same trend.

For prospective new nuclear plants, the fuel element is even less significant (see below). The typical front end nuclear fuel cost is typically only 15-20% of the total, as opposed to 30-40% for operating nuclear plants.

Understanding the cost of new generating capacity and its output requires careful analysis of what is in any set of figures. There are three broad components: capital, finance and operating costs. Capital and financing costs make up the project cost.

Capital costs comprise several things: the bare plant cost (usually identified as engineering-procurement-construction—EPC—cost), the owner's costs (land, cooling infrastructure, administration and associated buildings, site works, switchyards, project management, licences, etc), cost escalation and inflation. Owner's costs may include transmission infrastructure. The term "overnight capital cost" is often used, meaning EPC plus owners' costs and excluding financing, escalation due to increased material and labour costs, and inflation. Construction cost—sometimes called "all-in cost", adds to overnight cost any escalation and interest during construction and up to the start of construction. It is expressed in the same units as overnight cost and is useful for identifying the total cost of construction and for determining the effects of construction delays. In general the construction costs of nuclear power plants are significantly higher than for coal- or gas-fired plants because of the need to use special materials, and to incorporate sophisticated safety features and back-up control equipment. These contribute much of the nuclear generation cost, but once the plant is built the cost variables are minor.

Long construction periods will push up financing costs, and in the past they have done so spectacularly. In Asia construction times have tended to be shorter, for instance the new-generation 1300 MWe Japanese reactors which began operating in 1996 and 1997 were built in a little over four years, and 48 to 54 months is typical projection for plants today.

Decommissioning costs are about 9-15% of the initial capital cost of a nuclear power plant. But when discounted, they contribute only a few percent to the investment cost and even less to the generation cost. In the USA they account for 0.1-0.2 cent/kWh, which is no more than 5% of the cost of the electricity produced.

Financing costs will depend on the rate of interest on debt, the debt-equity ratio, and if it is regulated, how the capital costs are recovered. There must also be an allowance for a rate of return on equity, which is risk capital.

Operating costs include operating and maintenance (O&M) plus fuel. Fuel cost figures include used fuel management and final waste disposal. These costs, while usually external for other technologies, are internal for nuclear power (ie they have to be paid or set aside securely by the utility generating the power, and the cost passed on to the customer in the actual tariff).

This "back-end" of the fuel cycle, including used fuel storage or disposal in a waste repository, contributes up to 10% of the overall costs per kWh,—rather less if there is direct disposal of used fuel rather than reprocessing. The $26 billion US used fuel program is funded by a 0.1 cent/kWh levy.

Calculations of relative generating costs are made using levelised costs, meaning average costs of producing electricity including capital, finance, owner's costs on site, fuel and operation over a plant's lifetime, with provision for decommissioning and waste disposal.

It is important to note that capital cost figures quoted by reactor vendors, or which are general and not site-specific, will usually just be for EPC costs. This is because owner's costs will vary hugely, most of all according to whether a plant is Greenfield or at an established site, perhaps replacing an old plant.

Mid 2008 vendor figures for overnight costs (excluding owner's costs) have been quoted as:

GE-Hitachi ESBWR just under $3000/kW
GE-Hitachi ABWR just over $3000/kW
Westinghouse AP1000 about $3000/kW

There are several possible sources of variation which preclude confident comparison of overnight or EPC (Engineering, Procurement & Construction) capital costs—eg whether initial core load of fuel is included. Much more obvious is whether the price is for the nuclear island alone (Nuclear Steam Supply System) or the whole plant including turbines and generators—all the above figures include these. Further differences relate to site works such as cooling towers as well as land and permitting—usually they are all owner's costs as outlined earlier in this section. Financing costs are additional, adding typically around 30%, and finally there is the question of whether cost figures are in current (or specified year) dollar values.

www.ingramcontent.com/pod-product-compliance
Lightning Source LLC
Chambersburg PA
CBHW030941180526
45163CB00002B/653